EINSTEIN ON EINSTEIN

EINSTEIN ON EINSTEIN

AUTOBIOGRAPHICAL AND SCIENTIFIC REFLECTIONS

HANOCH GUTFREUND
& JÜRGEN RENN

PRINCETON UNIVERSITY PRESS
PRINCETON AND OXFORD

Requests for permission to reproduce material from this work should be sent to permissions@press.princeton.edu

Published by Princeton University Press
41 William Street, Princeton, New Jersey 08540
6 Oxford Street, Woodstock, Oxfordshire OX20 1TR

press.princeton.edu

All Rights Reserved
ISBN 9780691183602
ISBN (e-book) 9780691200118

Library of Congress Cataloging-in-Publication Data

Names: Gutfreund, Hanoch, author. | Renn, Jürgen, 1956- author. |
    Einstein, Albert, 1879–1955. Notas autobiográficas. English.
Title: Einstein on Einstein : autobiographical and scientific reflections /
    Hanoch Gutfreund and Jürgen Renn.
Description: Princeton : Princeton University Press, [2020] | Includes
    bibliographical references and index.
Identifiers: LCCN 2019041736 (print) | LCCN 2019041737 (ebook) |
    ISBN 9780691183602 (hardback) | ISBN 9780691200118 (ebook)
Subjects: LCSH: Einstein, Albert, 1879–1955. | Einstein, Albert, 1879–1955—Influence. |
    Physicists—Biography. | Physicists—Intellectual life.
Classification: LCC QC16.E5 G88 2020 (print) | LCC QC16.E5 (ebook) | DDC
    530.092—dc23
LC record available at https://lccn.loc.gov/2019041736
LC ebook record available at https://lccn.loc.gov/2019041737

British Library Cataloging-in-Publication Data is available

Editorial: Eric Crahan, Thalia Leaf
Production Editorial: Terri O'Prey
Production: Danielle Amatucci
Publicity: Sara Henning-Stout, Kate Farquhar-Thomson
Copyeditor: Beth Gianfagna

Jacket image: German-born theoretical physicist Albert Einstein (1879–1955) at home in Princeton, New Jersey, 1944. (Photo by Popperfoto / Getty Images)

This book has been composed in New Century Schoolbook, Minion Pro

Printed on acid-free paper. ∞

Printed in the United States of America

10 9 8 7 6 5 4 3 2 1

"IS THIS SUPPOSED TO BE AN OBITUARY?" THE ASTONISHED READER WILL LIKELY ASK. I WOULD LIKE TO REPLY: ESSENTIALLY YES. FOR THE ESSENTIAL IN THE BEING IN A MAN OF MY TYPE LIES PRECISELY IN *WHAT* HE THINKS AND *HOW* HE THINKS, NOT IN WHAT HE DOES OR SUFFERS. CONSEQUENTLY, THE OBITUARY CAN LIMIT ITSELF IN THE MAIN TO THE COMMUNICATING OF THOUGHTS THAT HAVE PLAYED A CONSIDERABLE ROLE IN MY ENDEAVORS.

(*Autobiographical Notes*, p. 31 [p. 165])

CONTENTS

INTRODUCTION

EACH OF OUR PREVIOUS BOOKS FEATURED A CANONICAL TEXT BY ALBERT EINSTEIN around which we built a whole narrative, placing that text in its historical and scientific context. Likewise, the present book features his *Autobiographical Notes*, published in *Albert Einstein: Philosopher Scientist*, volume 7 of The Library of Living Philosophers, a series initiated and directed by Paul A. Schilpp.[1] We have accompanied Einstein's short autobiographical account with interpretative essays that investigate from different angles its genesis, impact, and contexts, and we have supplemented it with additional historical documents. With this book, we hope to contribute to the accessibility and appreciation of Einstein's *Notes* as a canonical text of modern science and philosophy.

Einstein's *Autobiographical Notes* represents a key document of twentieth-century thought—one that especially illuminates the role of science in making the modern world. Like a focal lens, it collects the various thought traditions brought together in giving rise to the new physical world picture that was the result of the intellectual revolution initiated by Einstein and his peers. The essay offers a unique, introspective view of how this upheaval came about and thus constitutes a hitherto neglected counterpart to the autobiographies of politicians, writers, artists, or other actors who have played such an important role in revealing the subjective side of the tumultuous history of the twentieth century. Einstein's *Autobiographical Notes* is indeed an artistic document itself, taking the reader on an imaginative journey from the childhood to the last questions of the aging scientist. Even readers without any scientific background will be engrossed by Einstein's account of his dramatic life story.

The book is organized in six major parts. Part I offers some general historical background to the genesis of the *Autobiographical Notes* with a special focus on the origin and scope of Schilpp's monumental enterprise, The Library of Living Philosophers, and on historical developments in the year 1946, the year of writing these notes. This year marked a historical turning point: after the end of World War II, after the Holocaust, after the bombs dropped on Hiroshima and Nagasaki, and at the beginning of the Cold War. These developments had a profound effect on Einstein, on his public activities, on his writings, and on his mood. Although none of it is explicitly reflected in the *Autobiographical Notes*, it is worth remembering in what environment the memoir was written. In this preliminary part, we also compare Einstein's *Notes* with the *Scientific Autobiography* of Max Planck. The latter is one of many similar essays written by scientists and philosophers describing their intellectual odyssey.

Part II, the main part of the book and perhaps our most original contribution, begins with an introductory essay on the quest for a unified scientific worldview, followed by twelve essays commenting on the major themes of Einstein's text. Their goal is to unravel Einstein's convoluted narrative and to highlight his autobiographical reminiscences in their biographical context. We compare them with his perception of the different stages and chapters in his scientific life, as illustrated by his writings and correspondence at the time of their development. Conversely, we take a fresh look at his biography from the perspective of his autobiographical account.

Schilpp's volume on Einstein contains, after the *Autobiographical Notes*, twenty-five descriptive and critical essays by other physicists and philosophers commenting on his work. In Part III, we describe Schilpp's selection process and provide biographical information on the selected contributors. We then summarize and analyze Einstein's responses to their remarks. The latter supplement the *Notes* by shedding additional light on Einstein's scientific-philosophical worldview. We also quote from unpublished versions of his responses.

Part IV reproduces another remarkable text, Einstein's "Autobiographical Sketch," written about a month before his death in 1955. It is the first time that this text has been published in English. Einstein wrote it as his contribution to the one hundredth anniversary volume of the ETH in Zurich. He agreed to do it because this would give him an opportunity to express gratitude to his personal friend and scientific collaborator Marcel Grossmann. We discuss the contents and the context of this unique document.

Einstein was the only scientist included in Schilpp's Library of Living Philosophers. In Part V, the concluding section of our commentary, we attempt to show why the title *Philosopher-Scientist* is so appropriate.

Part VI presents a reprint of the English translation of the *Autobiographical Notes*. The work was written in German and translated into English by Schilpp with the help of Peter Bergmann, and Einstein approved the translation. In Schilpp's volume, the German and English texts were reproduced side by side.

In addition to representative historical illustrations and scans of selected passages from Einstein's handwritten manuscripts, we have chosen to illustrate the text with drawings by the artist Laurent Taudin. They accompany Einstein's reminiscences of his intellectual journey with an artistic, poetic walk as a metaphor of this journey, adding a light, yet thought-provoking, flavor. We appreciate Laurent's imaginative grasping of the human essence of Einstein's effort.

ACKNOWLEDGMENTS

This project owes a special debt to two institutions, which were directly and indirectly involved. The Hebrew University allowed us unlimited access and use of archival material, and the Max Planck Institute for the History of Science became the venue where this project was created. We are therefore grateful for the support of both these institutions.

One of us (H. G.) is grateful to the Max Planck Institute for the History of Science for its hospitality during numerous visits in the course of this work.

We are grateful to Dr. Roni Gross, director of the Albert Einstein Archives (AEA) for his assistance. His deputy, Mrs. Chaya Becker, deserves special thanks for her invaluable help with archival material at the AEA and with other sources.

We would like to acknowledge Sabine Bertram for her professional and effective assistance with the bibliographical research.

We are grateful to Nicholas L. Guardiano, research specialist from the Special Collections Research Center, Morris Library at Southern Illinois University, and to Paula McNally, an independent researcher, for their guidance and search for relevant archival material in the Schilpp archival collection.

Finally, we acknowledge with appreciation and gratitude the invaluable editorial assistance and professional support of Lindy Divarci.

NOTE

1. Paul Arthur Schilpp, ed., "Albert Einstein: Autobiographical Notes (in German, and in English Translation)," in *Albert Einstein: Philosopher-Scientist*, The Library of Living Philosophers, vol. 7 (Evanston, IL: Library of Living Philosophers, 1949), 1–96. The English translation from the separate bilingual edition of *Autobiographical Notes* is reproduced in this volume: *Albert Einstein: Autobiographical Notes*, ed. Paul Arthur Schilpp (La Salle, IL: Open Court, 1979).

PART I

PRELIMINARIES

1
THE GENESIS AND SCOPE OF THE *AUTOBIOGRAPHICAL NOTES*

THE MATTER WAS OF THE UTMOST IMPORTANCE, AT LEAST TO THE SENDER OF THE LETter. A turning point in world history seemed to be imminent. The message was to be heard all over America, by over 25 million listeners, and from there the message would surely spread across the globe. There was only one person who could authentically stress its urgency and lend it universal credibility, a man sixty-seven years of age, not a politician but a scientist. But given his fragile health, he needed to be convinced to take on the long journey from the East Coast to the Great Lakes, either by train or by plane and deliver a speech at a stadium in front of forty thousand people that would simultaneously be transmitted by radio.

The urgent issue was the establishment of a world government—in the aftermath of the most devastating war and the most horrific genocide the world had ever seen and on the eve of an even larger catastrophe casting its shadow on the fate of mankind as a whole. It was April 1946, about a year after the liberation of Auschwitz and the dropping of atomic bombs on Hiroshima and Nagasaki. While leading Nazi officials were sentenced in Nuremberg for their crimes against humanity, in March 1946 British Prime Minister Winston Churchill spoke for the first time of the Iron Curtain dividing Europe; later he would argue in favor of the United States of Europe. In the beginning of the year, the General Assembly of the United Nations had met for the first time; the first radar contact with the moon was established on the same day. The time seemed ripe for a planetary view on the future of mankind. Yet, as citizens turned their gaze from the horrors of the immediate past to the immediate future, a new abyss was opening, that of the Cold War, which would hover for much of the rest of the century at the brink of a hot war with the potential to destroy the entire planet.

That was one of the reasons why the establishment of a world government seemed to be so urgent to the sender of the letter: "What is absolutely necessary today—actually *sine qua non*—is world government! But of course it will not happen unless more and more people realize this, even if they are only driven to it through fear of the atomic bomb."[1] In May of the same year another letter insisted on the necessity to address the millions in favor of a world government "for the sake of humanity": "I am indeed absolutely certain that you could not support the cause that is dear to all our hearts—namely to save the world from annihilation—any better, most honorable Mr. Einstein, than by accepting the present invitation."[2]

The sender of the letters was Paul Arthur Schilpp, a professor of philosophy, of German origin, who had taught after World War I at several American universities and was now deeply involved in the Brotherhood Banquet, a mass movement organized by the National Conference of Christians and Jews in Chicago. The recipient was none other than Albert Einstein, the iconic scientist who had since long associated his fame in science with the cause to save mankind from the perils of the Atomic Age he himself had inadvertently helped to bring about, if only as a theoretician concerned with detached questions concerning the universe. The two men also had other business together. During the time they exchanged letters about how best to save the world, Schilpp suggested that Albert Einstein write an intellectual autobiography. Schilpp had initiated and edited a series of volumes titled The Library of Living Philosophers and expressed the wish to devote one volume to *Albert Einstein: Philosopher-Scientist*.

It took some time and required Schilpp's persuasive skills before Einstein agreed, in a letter to Schilpp, to deliver a handwritten scientific autobiography and a response to critical essays by a selected group of physicists and philosophers that were to be included in that volume.[3] Thus, at the age of sixty-seven, Einstein did what he had refused to do in the past—he sat down to write his *Autobiographical Notes*. Once he agreed, he acknowledged that ". . . it is a good thing to show those who are striving alongside of us how our own striving and searching appears in retrospect." At the same time, he warned the reader, "Every reminiscence is colored by one's present state, hence by a deceptive point of view" (p. 3 [p. 157]). This did not deter him from undertaking this project, because only he had access to his conscious experience to share it with others.

How were the two enterprises that united Schilpp and Einstein connected? And how would Einstein describe his own striving? Would he refer to the global crisis of the world and take the occasion to write an account in the style of Mahatma Gandhi, whose life he considered one of the greatest testimonies of true human greatness? Gandhi's autobiography, *The Story of My Experiments with Truth*, focuses on the quest for a spiritual and moral life that, in the midst of the turmoil of the world, offered him the wisdom and strength for political protest. Gandhi's autobiography is not just the story of an inner journey but the realistic portrait of a troubled world. Which are the conflicts, temptations, and aspirations that would take center stage in Einstein's autobiography?

Schilpp's aspirations were, in any case, satisfied. When he received the manuscript of the *Notes*, he responded enthusiastically:

Thus, honorable Professor Einstein: my best and most sincere thanks! And not just *my* thanks (because who am I?), but rather in this case I may already extend to you the most profound gratitude of innumerable readers and even of those as yet unborn people who in the coming decades—and yes, even centuries—will be grateful to you for this marvelous (and altogether Einsteinian) work and will owe a debt of gratitude. This you have done simply splendidly! If I, after having read your autobiography, think back to when I first asked you to contribute to the creation of such a volume, and you said "No!," and that the entire world could have been deprived of and remained without this wonderful autobiography, I still shudder at the thought.[4]

Every autobiography is a time machine—of a kind that relativity theory has not accounted for and never will. It draws the reader into the world of another mind of another time, it draws the author into his or her own past, and it speaks to all those fellow travelers in time who have undertaken a similar journey—or will do so in the future. Einstein's autobiography carries our imagination to the world immediately after World War II, to the small university town of Princeton, New Jersey, to a modest house on Mercer Street where Einstein sat down, writing his own obituary as he mockingly began his text. His own thoughts quickly escaped from this world, however, to a time before the great wars, to the youth he spent in Germany, Switzerland, and Italy. These were troubled times as well, even if seemingly still infinitely far from the catastrophes of the twentieth century. But even the usual troubles of the world, the struggle to make a living, the political tensions, the foreshadowing of the future drama, all fade into the background of Einstein's account.

Central to Einstein's autobiography are the troubles, challenges, and tensions encountered along his quest for a scientific worldview. Throughout the entire text it is clear that what counts at the end is the striving and struggle and not the final formulation of successful breakthroughs, which brought Einstein universal fame. He does not even mention the groundbreaking papers of the "miraculous year" 1905, on light quanta, on Brownian motion, and on special relativity, which constitute his Copernican revolution and became the pillars of modern physics.[5] On the other hand, he explores the origins of those achievements, his thought process, and his search for new principles. Likewise, he does not mention his final formulation of the general theory of relativity in November 1915, which was celebrated as another great revolution one hundred years later. This theory became the basis of modern cosmology and, hence, of our understanding of the universe. His recollections on the emergence of the theory of general relativity rather focus on why it took seven more years from the seminal idea to its groundbreaking consequences.

Einstein's quest, as he describes it in his autobiography, was, at the same time, the search for the role and path of a young man curious about his own as well as humanity's place within the world. On this account, finding one's place in the world and comprehending its inner secrets become part of the same quest. Einstein is known for his often ironically distanced way of treating God as an interlocutor and counterpart of his scientific quest. Einstein did not believe in the monotheistic religions' conception of God's role in punishing and rewarding human beings. In this sense he objected to the concept of a "personal god." For him, God was an embodiment of the laws and harmony of nature, and it is with this god that he maintained a lifelong dialogue. On this we can quote a characteristic statement: "I believe in Spinoza's God who reveals himself in the harmony of all that exists, but not in a God who concerns himself with the fate and actions of human beings."[6] "Subtle is the Lord, but Malicious is He not" he would claim, or show himself certain that "God does not play dice."[7] Einstein's *Autobiographical Notes* sometimes addresses the reader, but it actually lets the reader participate in how his dialogue with God—that is, his struggle for an understanding of the physical world—evolved over time.[8]

In a word, Einstein's *Autobiographical Notes* is, in a sense, his confession, a secular counterpart to the famous *Confessions* of St. Augustine, a monument of Western thought. Augustine of Hippo also lived in a time of great changes, in the fourth and early fifth centuries, which also saw a division of East and West and the beginning of the decline of the western Roman Empire. As it turned out, Einstein had read the *Confessions* and was apparently fascinated by the way St. Augustine accounted for his inner journey within a troubled world. It is not an exaggeration to claim that this text of late antiquity acted at a distance of over a millennium and a half, shaping the way Einstein presented his own life to himself and his readership. It can be characterized as a striving for inner freedom and comfort within the larger community of those striving for an eternal truth that will always be in flux.

This is just one of the surprising insights Einstein's extraordinary book has to offer, which is indeed perhaps the most extraordinary of all books he wrote. Here we undertake the attempt of a new reading of this text. Based the background of decades of Einstein scholarship, we are now in a better position to place this text within his own biography and its manifold contexts, to understand the allusions he makes, to interpret the omissions, and to grasp the subtle hints he gives. Just like St. Augustine's text, Einstein's *Autobiographical Notes* is a message in a bottle, a time capsule coming from a specific place and historical situation, but conveying insights that by far transcend those specifics, insights that capture the essence of a lifetime of thinking about the universe and humanity's place within it.

Einstein's narrative is confined to the early and the late stages of his scientific career. There is hardly anything about his activities after his formative years as a scientist and before his later years in Princeton. The emphasis is on his work during the years preceding 1905 and on his road to the general theory of relativity. From there he jumps directly to his concerns about the status of quantum mechanics and the quest for a unified field theory at the time of writing these notes.

The main part of our book consists of thirteen commentary chapters. The first of these introduces the quest for a unified worldview as a theme on the agenda of the scientific community at the beginning of the twentieth century. The other twelve chapters essentially trace Einstein's text. The second chapter describes how his personality and his chosen life course evolved from his childhood years and social environment, mentioning specifically two biographical experiences—his brief religious episode at the age of twelve and his formal school education. In the third chapter we discuss Einstein's introspective account of his way of thinking, specifically, thinking that leads to scientific discovery. Einstein believed that scientific inquiry should be based on and guided by epistemological principles. This belief motivated him to formulate an "epistemological credo," which we discuss in our commentary. The next two chapters are devoted to an exposition of classical nineteenth-century physics and its drawbacks, leading to its final decline with Einstein's revolutionary discoveries.

The following chapters deal with Einstein's work leading to the *annus mirabilis* 1905. The discussion of this period in the *Notes* is brief, narrated in an entangled style and sometimes confined to mere hints. We begin with his reaction to the groundbreaking work of Max Planck on black-body radiation. We then discuss Einstein's own derivation of statistical mechanics and explore his motivation to undertake this endeavor, comparing his formulation with the classical kinetic theory of thermodynamics developed by

Ludwig Boltzmann. The next chapter is devoted to Einstein's interest in thermodynamic fluctuations, which led to the understanding of Brownian motion and, eventually, to convincing evidence for the reality of atoms. We then discuss Einstein's thought experiment on a reflecting mirror suspended in the radiation field enclosed in a cavity. The analysis of this setup provided compelling arguments for the corpuscular nature of light. Another chapter, related to Einstein's work in the period preceding the "miraculous year," is devoted to the origins of the special theory of relativity, which emerged from the same framework of considerations that led to his other achievements in the year 1905.

Not much is known about this period from contemporary documents, beyond the papers published in those years. There are sporadic references to Einstein's ideas and interests during these years in the love letters he exchanged with his student companion and wife-to-be, Mileva Marić.[9] We shall quote them whenever appropriate. The *Autobiographical Notes* and these love letters provide two complementary perspectives: one from the vista point of old age and one from the midst of the struggles.

We then comment on Einstein's account of his road to general relativity and on the difficulties he encountered as he struggled toward this goal. We have written an entire book on this subject.[10] Here we compare Einstein's recollections of this process with the perception based on extensive contemporary documents and correspondence. Einstein's answer to his question "Why were another seven years required?" (actually it took eight years) is not the full story. We can wonder if this is how he remembered it, or if this is how he wished it to be remembered.

Further chapters are devoted to Einstein's views on the status and future of quantum mechanics and to his search for a unified field theory and his opinion on the preferred approach to this goal. Here it is not a question of reminiscences of the past. Surprisingly, he does not at all refer to his famous debates with Niels Bohr on quantum mechanics in the 1920s nor to the different approaches toward a unified field theory that he explored in the 1920s and 1930s. He focuses exclusively on his views on these two topics at the time of writing his *Autobiographical Notes*, evidently using this work as a conduit for documenting what he considered an important part of his scientific legacy. His attempt to formulate a unified field theory begins with the quest for a new and broader symmetry. This reflects Einstein's enduring legacy about the role of symmetry in physics. Symmetry comes first, and it determines the laws of physics and the corresponding equations. He had applied this principle, previously, in his discussion of prerelativity physics, of special relativity, and of general relativity, and he now applies it in his search for a unified field theory.[11]

At this stage in his life, Einstein believed that the most promising path toward a unified theory was based on the assumption of nonsymmetric fields, and the last part of the *Notes* presents a brief description of this approach. He had devoted the last ten years of his life, exclusively, to the exploration of this option. With Einstein, we conclude our commentaries, pointing out how his lifelong odyssey led him to the conclusion that the future of physics lies in a generalization of his theory of gravity, still based on the classical notion of continuous fields even though he was open to considering alternative foundations of physics.

The inclusion of Einstein in The Library of Living Philosophers and the meaningful title of the specific volume—*Albert Einstein: Philosopher-Scientist*—shed a focused light on the philosophical and epistemological thinking that accompanied his scientific journey in search of a unified worldview. This is clearly demonstrated in his own account of this journey in the *Autobiographical Notes* and is further expanded and amplified in his responses to the critical essays by selected philosophers and physicists included in this volume. Our account of these responses constitutes another major part of this book. We justify the "philosopher-scientist" attribute of Einstein in our concluding remarks.

NOTES

1. Schilpp to Einstein, 4 April 1946, Albert Einstein Archives (hereafter AEA) 80-507. Unless otherwise indicated, German texts have been translated into English by the authors.
2. Schilpp to Einstein, 3 May 1946, AEA 80-508.
3. Einstein to Schilpp, 29 May 1946, AEA 42-513.
4. Schilpp to Einstein, 8 February 1947, AEA 42-515.
5. See, for example, Jürgen Renn and Robert Rynasiewicz, "Einstein's Copernican Revolution," in *The Cambridge Companion to Einstein*, ed. Michel Janssen and Christoph Lehner (Cambridge: Cambridge University Press, 2014).
6. Alice Calaprice, ed., *The New Quotable Einstein*, rev. ed. (Princeton, NJ: Princeton University Press, 2005), 197.
7. "Raffiniert ist der Herrgott, aber boshaft ist er nicht," CPAE vol. 12, p. liii. "Jedenfalls bin ich überzeugt, daß der nicht würfelt," Albert Einstein to Max Born, 4 December 1926, in Albert Einstein and Max Born, *Briefwechsel, 1916–1955* (Munich: Nymphenburger, 1969), 129–130.
8. For Einstein's concept of God, see Yehuda Elkana, "Einstein and God," in *Einstein for the 21st Century: His Legacy in Science, Art, and Modern Culture*, ed. Peter L. Galison, Gerald Holton, and Silvan S. Schweber (Princeton, NJ: Princeton University Press, 2008), 35–47.
9. See Jürgen Renn and Robert Schulmann, eds., *Albert Einstein—Mileva Marić: The Love Letters* (Princeton, NJ: Princeton University Press, 1992).
10. Hanoch Gutfreund and Jürgen Renn, *The Road to Relativity: The History and Meaning of Einstein's "The Foundation of General Relativity," Featuring the Original Manuscript of Einstein's Masterpiece* (Princeton, NJ: Princeton University Press, 2015).
11. This point is discussed in Albert Einstein, *The Meaning of Relativity*, 5th ed. (Princeton, NJ: Princeton University Press, 1955), 1–23; see also Hanoch Gutfreund and Jürgen Renn, *The Formative Years of Relativity: The History and Meaning of Einstein's Princeton Lectures* (Princeton, NJ: Princeton University Press, 2017), 7.

2
SCHILPP'S ENTERPRISE
THE LIBRARY OF LIVING PHILOSOPHERS

THE LIBRARY OF LIVING PHILOSOPHERS (LLP) IS A SERIES OF VOLUMES CONCEIVED AND initiated by Paul Arthur Schilpp (1897–1993). Schilpp was born in Germany and emigrated before World War I to the United States, where he taught philosophy at several universities. He was a political activist whose ideological principles and commitments were compatible with those of Einstein.

Every one of the LLP volumes is devoted to a single living philosopher. The main goal of the library was to provide a platform for a philosopher to reply to his or her interpreters and critics while still alive, hopefully resolving disputes about what someone really meant. This may have been a naive expectation, because the philosopher's reply may be the subject of different interpretations just as much as the original writings. Nevertheless, the LLP has become an important philosophical resource.

Arthur Schilpp with Albert Einstein in the latter's study in Princeton, 28 December 1947. With permission of the American Institute of Physics.

In line with the primary goal of LLP, each volume contains the following four parts:

- an intellectual autobiography of the philosopher, whenever this can be secured,
- a series of expository and critical articles written by the leading exponents and opponents of the philosopher's thought,
- a reply to the critics and commentators by the philosopher,
- a bibliography of the writings of the philosopher to provide a ready instrument to give access to his or her publications.

In a seminar offered to graduate students at the philosophy department of Southern Illinois University in 1967, Schilpp described how the LLP came into existence.[1] He recalled that in 1933, as chairman of the philosophy department at the University of the Pacific in Stockton, California, he attended a lecture by the German-British philosopher and proponent of American pragmatism, Ferdinand Canning Scott Schiller. The title of the lecture was "Must Philosophers Disagree?" A few quotations from that lecture indicate how something akin to what later became The Library of Living Philosophers was born, at least as an idea in Schilpp's mind:

> The philosophic public is not inquisitive enough. By a sedate (or professional) convention it does not ask philosophers what they mean, or why on earth they have written as they have, while they are alive. It waits till they are dead, and can no longer explain themselves, and then it starts guessing their riddles. Thereby it makes hay of them; it turns them into desiccated lecture-fodder, which provides innocuous sustenance for ruminant professors. . . . [T]hese can now speculate, safely, endlessly, and fruitlessly, about what a philosopher may have meant, nay must have meant; they are no longer in danger of being upset by his telling them what he *did* mean. . . . A further bar to fruitful discussion in philosophy is the curious etiquette which apparently taboos the asking of questions about a philosopher's meaning while he is still alive. . . . [T]his has certainly preserved the vitality of many insoluble questions and interminable controversies which fill the histories of philosophy, and which could have been ended at once by asking the living philosophers a few searching questions. . . . [T]heoretically, they *could* be discussed, openly, profitably and effectively, and settled to a large extent.[2]

Schilpp had no doubt about the truth of Schiller's remarks. He could not understand why nothing had ever been done about it by anyone. He was inspired to remedy this situation. The idea itself was clear enough: give a great philosopher an opportunity to explain himself or herself further and to reply to both disciples and critics while he or she is still alive.

Looking back at his experience after having edited the first twelve volumes of LLP, Schilpp realized that Schiller was too optimistic and that it was impossible to end "the interminable controversies which fill the histories of philosophy . . . by asking the living philosophers a few searching questions." His experience with LLP had demonstrated to him this impossibility beyond any question. Philipp Frank, one of the first to be invited to contribute to the Einstein volume, warned him on this point: "Your general idea to ask a living philosopher bluntly what he meant to say seems to me an excellent one. There is, however, a flaw in it. For it supposes that the 'living philosophers' are able to say clearly what they have meant. Unfortunately, the language of the living philosophers is not easier to understand than the books of deceased philosophers. However, you give to the living a last opportunity to be clear which he may take advantage of as long as it is time."[3] Schilpp confessed that he was not sure if the series of LLP volumes would ever have come into existence if he had known in the 1930s what he knew when he delivered his 1967 seminar.

Schilpp began to implement his dream four years after the Schiller lecture, when he moved to Northwestern University. In his "Glimpses of a Personal History," Schilpp describes his attempts to secure funding for this project.[4] The president of the university showed interest in the project, though in a less ambitious format. He instructed his director of development to explore funding options. After a year of unsuccessful attempts, Schilpp was told that it was ". . . easier to get five million dollars for cancer research than fifty cents for philosophy." Schilpp approached different foundations and received minimal grants for planning but not for publication. The publication of the first volumes was enabled by a personal loan from the university administration. It was not until the Einstein volume (no. 7) was on sale that Schilpp was able to pay off the publication costs of the books to the university. The printing of this volume was doubled to five thousand copies, all of which were sold within ten months of publication. This success was a turning point in the publication history of the LLP, which turned into a legal entity in the autumn of 1950.

Schilpp served as the chief editor of LLP between 1939 and 1981. The series continued after that with a new editorial board. Thirty-five volumes have been published so far (see box). Albert Einstein remains the only author who is not predominantly a philosopher.

THE LIBRARY OF LIVING PHILOSOPHERS (35 VOLUMES)

John Dewey (1939)
George Santayana (1940)
Alfred North Whitehead (1941)
G. E. Moore (1942)
Bertrand Russell (1944)
Ernst Cassirer (1949)
Albert Einstein (1949)
Sarvepalli Radhakrishnan (1952)
Karl Jaspers (1957)
C. D. Broad (1959)
Rudolf Carnap (1963)
Martin Buber (1967)
C. I. Lewis (1968)
Karl Popper (1974)
Brand Blanshard (1980)
Jean-Paul Sartre (1981)
Gabriel Marcel (1984)
W. V. Quine (1986)

Georg Henrik von Wright (1989)
Charles Hartshorne (1991)
A. J. Ayer (1992)
Paul Ricoeur (1995)
Paul Weiss (1995)
Hans-Georg Gadamer (1996)
Robert Chisholm (1998)
P. F. Strawson (1998)
Donald Davidson (1999)
Seyyed Hossein Nasr (2001)
Marjorie Grene (2002)
Jaakko Hintikka (2006)
Michael Dummett (2007)
Richard Rorty (2009)
Arthur C. Danto (2013)
Hilary Putnam (2015)
Umberto Eco (2017)

NOTES

1. Paul A. Schilpp, "Glimpses of a Personal History," Paul Arthur Schilpp papers, Special Collections Research Center, Morris Library, Southern Illinois University Carbondale, box 21, folder 2.

2. F.S.C. Schiller, *Must Philosophers Disagree? and Other Essays in Popular Philosophy* (London: Macmillan, 1934), 11, 13, 14.

3. Philipp Frank to Schilpp, 10 February 1946, Paul Arthur Schilpp papers, Special Collections Research Center, Morris Library, Southern Illinois University Carbondale, box 12, folder 18.

4. Schilpp, "Glimpses of a Personal History."

3
HISTORICAL BACKGROUND

THE YEAR 1946

Stormy weather—1946,
a year of *vita activa*.

EINSTEIN CONFIRMED HIS AGREEMENT TO WRITE THE *AUTOBIOGRAPHICAL NOTES* FOR the planned volume devoted to his work at the end of May 1946.[1] Six months later, Schilpp acknowledged the message from Einstein's secretary, Helen Dukas, that Einstein's scientific autobiography was ready.[2] Thus, the *Autobiographical Notes* were conceived and completed throughout 1946. The year 1946 was one of the most active years in Einstein's public life and political activities. Our brief account of these activities is based on the material and commentaries in Nathan and Norden's *Einstein on Peace* (1960), Rowe and Schulmann's *Einstein on Politics* (2007), Jerome and Taylor's *Einstein on Race and Racism* (2005, chap. 8), as well as on the Einstein-Schilpp correspondence.

THE IDEAL OF A WORLD GOVERNMENT

In January 1946, the magazine *Survey Graphic* organized a symposium under the title "Year 1: Atomic Age." It featured Einstein's statement: "The weapons of modern warfare have developed to such a degree that, in another world war, the victor would probably suffer not much less than the vanquished. . . . I firmly believe that the majority of peoples in the world would prefer to live in peace and security rather than have their particular nation pursue a policy of unrestricted national sovereignty. Mankind's desire for peace can be realized only by the creation of world government."[3]

Thus, when Einstein received the two letters from Schilpp, cited in the introduction, urging him to convey to the general public his conviction about the necessity of a world government, he had already been deeply involved in promoting this idea. About a year earlier, he received from the writer and publisher of Hungarian origin, Emery Reves, a copy of the latter's recently published book *The Anatomy of Peace* (1945). The author argued that the United Nations Security Council was not adequate to secure peace because it was an instrument of power. The only way to prevent war would be world federalism, namely, a world government and a world law. Einstein embraced these ideas, and the next edition of *The Anatomy of Peace* appeared with his endorsement.

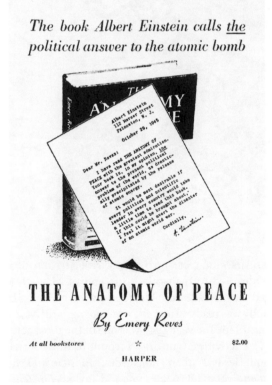

Advertisement for *The Anatomy of Peace* by Emery Reves, with Einstein's endorsement.

Einstein contributed to the broad popularity of these ideas and of Reves's book on many platforms and occasions. A widely read article was published in the November 1947 issue of *The Atlantic*, "Atomic War or Peace," based on a radio interview with Raymond Swing, himself an advocate of world government.

In May 1946, Einstein was appointed chairman of the Emergency Committee of Atomic Scientists, which served as a platform for promoting the idea of an international framework, like the United Nations, for the control of nuclear energy. With his appointment, Einstein began to campaign toward this goal, also keeping in mind the notion of a world government. In an interview published in the *New York Times* (23 June 1946), he said:

> Today the atomic bomb has altered profoundly the nature of the world as we knew it, and the human race consequently finds itself in a new habitat to which it must adapt its thinking. In the light of new knowledge, a world authority and an eventual world state are not just *desirable* in the name of brotherhood, they are *necessary* for survival. . . . Today we must abandon competition and secure cooperation. This must be the central fact in all our considerations of international affairs; otherwise we face certain disaster. Past thinking and methods did not prevent world wars. Future thinking *must* prevent wars.[4]

Einstein's devotion to the notion of world government is echoed in a letter to his lifelong friend Michele Besso, with whom he discussed physics and also shared his personal experiences, hopes, and frustrations: "If you occasionally hear my name mentioned in connection with political excursions, don't think that I spend much time on such things, as it would be sad to waste much energy on the meager soil of politics. From time to time, however the moment arrives when I cannot help myself, for instance, when one can draw the public's attention to *the necessity of world government* [italics added] without which all our human grandeur will go to the dogs."[5]

Einstein's motive to promote the idea of world government as a safeguard against the dangers of the "weapons of modern warfare" may have been his reaction to the public image of his connection with these weapons as the "grandfather of the bomb," reinforced by mainstream media. The 1 July 1946 issue of *Time* magazine featured, on its cover page, Einstein against the background of a mushroom cloud, which became a symbolic representation of the atom bomb. The caption under the picture reads: "Cosmoclast Einstein. All matter is speed and flame." It is true that Einstein's formula $E = mc^2$ tells us that a small amount of mass can be transformed into a huge amount of energy that can even destroy a city. But, the road from this formula to the atom bomb still required the work of thousands of technicians, engineers, and scientists. Einstein was not involved in this effort; he was not part of the Manhattan Project, and the participants in that project were not even allowed to discuss the matter with him. Yet his having been connected to the genesis of the atomic age haunted him for the rest of his life.

This was Einstein's mood and mindset when Schilpp asked him to address a mass peace rally on the eve of Memorial Day (29 May 1946) in the Chicago Stadium, to be held by the organization of Students for Federal World Government. In preparation for this event, three students visited Einstein in Princeton and posed questions to be discussed at the rally. One reads: "Professor Einstein, what, precisely, is the real difference between World Government and the United Nations Organization? Would you care to enlighten us on this point?" Einstein gave a handwritten answer in German.

Einstein was unable to attend the rally in person, but he agreed to address it by radio. The American Broadcasting Company made it accessible to the listeners of its stations from coast to coast. The broadcast was conducted in form of an interview. We have the full text of this interview. In that interview, Einstein responded to the above question in English: "Under a world government I understand an institution whose decisions and prescriptions are binding for the individual states. It is an institution, therefore, which among the present nations of the world is analogous to the relationship which exists between the government in Washington D.C. and the 48 states of the Union. In its present form the UN does not possess the powers of a world government, because its decisions and determinations have no binding power over the individual governments."[6]

Schilpp introduced Einstein and added a question of his own: "Do you believe that mutual envy and hatred among nations can be overcome by a world government?" Einstein responded: "No, of course not. But a world government would be able to prevent that such emotional reactions should lead to acts of violence between nations." The main point emphasized in Einstein's address was that the solution to the problem of international peace "is linked solely to an agreement on a grand scale between this country and Russia." The whole event and, specifically, Einstein's address were reported extensively in the US press.

RACIAL DISCRIMINATION

In 1946, Einstein became also a civil rights activist. At the end of the war about one million black American soldiers returned home after fighting for freedom and democracy in Europe. They expected that their participation in the victory over fascism abroad would end their second-class citizenship and discrimination at home. Instead, they found themselves

U. S. AND RUSSIA HOLD PEACE KEY, EINSTEIN WARNS

Mathematician Urges Grand Scale Pact

The solution for world peace depends, ultimately upon agreement between the United States and Russia, Dr. Albert Einstein, noted mathematician, declared last night.

He spoke by radio from his home in Princeton, N. J., to a mass meeting of 5,000 students and adults in the Chicago Stadium sponsored by the Students for Federal World Government. The organization was founded by a group of war veterans studying at Northwestern university.

In his remarks on Russia and world peace, Dr. Einstein inferentially rapped his adopted country, the United States, for what he described as Russia's distress.

Sees Others Following

"It is no exaggeration to say that the solution of the real problem [world peace] is linked solely to an agreement on a grand scale between this country and Russia," he said, explaining that the two countries alone would be powerful enough to make other countries follow their bidding.

Dr. Einstein asserted that if a fundamental agreement with Russia appears impossible now the fault in part is that the United States has not made a serious attempt in that direction.

"There was no need," he went on, "to accept Fascist Argentina in the United Nations. There was no need to manufacture new atomic bombs without letup; nor was it necessary to delay proposed measures against Franco Spain. Russia's distress is a distress to whose origin we have contributed no little."

Thomas Among Speakers

Among other speakers were Norman Thomas, Socialist leader; Ely Culbertson, bridge expert; Clifton Fadiman, radio figure, and Sen. Taylor [D., Ida.]. Thomas assailed imperialism; Culbertson hit the U. N. as "an assembly of mice presided over by cats fighting like fury."

The collection hat was passed to help meet the student group's expenses to date, estimated at $9,500. This included $5,000 for the Stadium rent, which was paid. The collection was made, a press aid said, to help meet the rest of the obligations.

Chicago Tribune,
30 May 1946.

facing a newly hostile American population who resented these soldiers in uniform for having the audacity to consider themselves equal. Racial segregation was the rule in most of the United States, with separate and unequal public schools, buses, and beaches. This was the situation throughout the South, but also in New Jersey, where Einstein lived.

A wave of anti-black violence began in 1946, resulting in the death of fifty-six African Americans nationwide, mostly veterans. One of the most publicized instances of white resistance to black notions of equality forged in World War II occurred in February 1946, when five hundred Tennessee state troopers with submachine guns surrounded the African American community of Columbia, Tennessee. More than one hundred black men were arrested. Twenty-seven were charged with rioting and attempted murder, and two were shot awaiting bail in the local jail. The riot made national headlines. Einstein publicly joined the National Committee for Justice in Columbia, Tennessee, which was

Einstein addressing students at Lincoln University in Pennsylvania, 3 May 1946. John W. Mosley Photograph Collection, Temple University Libraries, Philadelphia.

headed by Eleanor Roosevelt. A few weeks later, on 3 May, he went to Lincoln University in Pennsylvania to speak to students and to receive an honorary degree. At this stage in his life he avoided public appearances at universities, but in this case made an exception. In his commencement speech, he said: "My trip to this institution was on behalf of a worthwhile cause. There is separation of colored people from white people in the United States. That separation is not a disease of colored people. It is a disease of white people. I do not intend to be quiet about it."[7]

Actually, Einstein began his civil rights activism earlier that year when he published in *Pageant* magazine "A Message to My Adopted Country," where he unequivocally states: "There is, however, a somber point in the social outlook of Americans. Their sense of equality and human dignity is mainly limited to men of white skins. Even among these there are prejudices of which I as a Jew am dearly conscious; but they are unimportant in comparison with the attitude of 'Whites' toward their fellow-citizens of darker complexion, particularly toward Negroes. . . . The more I feel an American, the more this situation pains me. I can escape the feeling of complicity in it only by speaking out."[8]

THE PALESTINE PROBLEM

In the aftermath of World War II and the Holocaust, the Zionist demand for the free entry of Jews into Palestine became an issue that attracted broad international attention. The British policy, yielding to Arab pressures, imposed severe restrictions on such immigration.

Albert Einstein surrounded by friends—Meyer Weisgal (*holding his arm*), Helen Dukas, and others—on the way to the hearing on Palestine of the Anglo-American Committee of Inquiry, Washington, DC, 11 January 1946. Photo by Alexander Archer.

In order to involve the United States in its—apparently unsolvable—Palestinian problem, Britain joined with the Americans to form the Anglo-American Committee of Inquiry to investigate European Jewish immigration and settlement in Palestine. Einstein was invited to testify before the committee in January 1946.[9] He used this opportunity to launch an attack on British colonial policy, which in his opinion made that nation unfit for further administration of its mandate over Palestine. He argued that the great majority of Jewish refugees in Europe should be settled in Palestine and expressed strong support for unlimited Jewish immigration into Palestine. But, to the dismay of his fellow Zionists, he dismissed the goal of a Jewish state. The final remark in his testimony reads: "The state idea is not according to my heart. I cannot understand why it is needed. It is connected with many difficulties and a narrow-mindedness. I believe it is bad." The Palestine issue was on Einstein's mind and is reflected in his correspondence throughout 1946 and beyond.

EINSTEIN'S PUBLIC ACTIVITIES VERSUS CONTEMPLATIVE WRITING

There is a stark contrast between the content and style of the *Autobiographical Notes* and Einstein's public life, evidenced by his multifarious activities in 1946. He was leading an intense *vita activa* and yet, at the same time praising a *vita contemplativa* in the *Notes*. There he did not mention his political activism at all. Even in private communication, as in the letter to Besso quoted above, he tended to downplay this aspect of his life. He writes in the *Notes*, "For the essential in the being of a man of my type lies precisely in *what* he thinks and *how* he thinks, not in what he does or suffers" (p. 31 [p. 165]). Yet, his political views and activities are not just add-ons to a life for science. They are evidently driven by the same inner urge as his quest for scientific knowledge. On numerous occasions, Einstein admits that he cannot help engaging himself—for example, in support of a world government or the cause of black people.

How are these two quests related? At first glance, one might think that their common root is the search for unity in a worldview, be it the unity of science on the basis of a unified field theory or the unity of mankind held together by a world government. But this may just be a superficial parallelism and too simple a reconciliation that hides a deeper tension at the root of his personality. Einstein himself was aware of this deeper tension. In 1931, he wrote an essay, "The World as I See It," in which he refers to a broad array of themes, ranging from science to art, religion, the ideal political order, and the meaning of life, which is presented in greater detail on p. 29.[10] In fact, this text may be considered as an earlier version of his "confessions." But in contrast to the *Autobiographical Notes*, there he explicitly addresses both his *vita contemplativa* and his *vita activa*. In all of these spheres of human life, Einstein held on to some basic principles that provided him with guidance in his life. Yet he realized that he himself was fundamentally torn between his inner life and sense of independence and his passion for human affairs, as expressed in his well-known statement: "My passionate sense of social justice and social responsibility has always contrasted oddly with my pronounced freedom from the need for direct contact with other human beings and human communities."[11]

It may have been a coincidence that the request to write his scientific autobiography came specifically in that particularly active year, 1946. But once it happened, it symbolizes, together with everything else that occurred in that year, some of the inner dynamics driving him in all the realms of human endeavor.

NOTES

1. Einstein to Schilpp, 29 May 1946, AEA 42-513.
2. Schilpp to Einstein, 7 December 1946, AEA 80-511.
3. Cited in Otto Nathan and Heinz Norden, eds., *Einstein on Peace* (New York: Avenel Books, 1960), chap. 12.
4. Reprinted in David E. Rowe and Robert Schulmann, *Einstein on Politics: His Private Thoughts and Public Stands on Nationalism, Zionism, War, Peace, and the Bomb* (Princeton, NJ: Princeton University Press, 2007), 383–388, here 383.
5. Einstein to Besso, 21 April 1946, AEA 7-381; cited in Rowe and Schulmann, *Einstein on Politics*, 345.
6. The full text of this interview is "Proceedings. Memorial Eve Rally, The Master Reporting Company Inc.," 29 May 1946, AEA 90-570.
7. *New York Times*, 4 May 1946, 7.
8. January 1946; reprinted in Rowe and Schulmann, *Einstein on Politics*, 474.
9. Einstein's testimony is reprinted in Rowe and Schulmann, *Einstein on Politics*, 340–344.
10. Albert Einstein, *Ideas and Opinions: Based on "Mein Weltbild,"* ed. Carl Seelig (New York: Bonanza Books, 1954), 8–11. See Part II, chapter 1, p. 28–29.
11. Albert Einstein, *The World as I See It*, trans. Alan Harris (New York: Citadel Press 2000), 2–3.

4
EINSTEIN'S *AUTOBIOGRAPHICAL NOTES* AND PLANCK'S *SCIENTIFIC AUTOBIOGRAPHY*

People complain that our generation has no philosophers. Quite unjustly: it is merely that today's philosophers sit in another department, their names are Planck and Einstein.

—Adolf von Harnack, quoted by Arnold Sommerfeld in Schilpp, *Albert Einstein: Philosopher-Scientist*, p. 99

THE GENRE OF INTELLECTUAL AUTOBIOGRAPHY HAS A LONG TRADITION. THERE IS, however, perhaps no more striking parallel to Einstein's memoir than Planck's autobiography.[1] Einstein and Planck were both the uncontested heroes in the transition from classical to modern physics. In many respects they were as different as two people, born and raised in similar environments, can be—separated not only by a deep chasm in political matters but also very dissimilar in every aspect of their private lives. Their relationship was rather complex, yet, over the years (they were together in Berlin between 1914 and the beginning of the Nazi regime), they developed mutual respect, collegiality, and friendship, which were stronger than the differences in their personalities and their diametrically opposed outlooks and actions in the social, national, and political arena. Their scientific worldviews were strikingly similar. They both foregrounded their intellectual efforts rather than their personal life histories. They both conceived physics as part of an intellectual quest for a comprehensive worldview. In short, they were both philosopher-scientists. In addition, their mutual esteem survived the hardships of their times.

In August 1944, Emil Abderhalden, president of the German Academy for Natural Science, Leopoldina, appealed to prominent members of the academy, urging them to write autobiographies, contributing by that to the history of the development of natural

In its inaugural year, the
Deutsche Physikalische Ge-
sellschaft (German Physical
Society) awarded the Max
Planck Medal to Albert
Einstein, 29 June 1929. Max
Planck himself, who was
the other recipient of the
award that year, is present-
ing the medal to Einstein in
Berlin. From the Institute
for Advanced Studies.

sciences. To Max Planck, he addressed the same request in a personal letter. Planck's first
reaction was negative, but following another appeal in December, he agreed that it was
his duty to history and to his community. At that time, at the age of eighty-six, Planck's
health was deteriorating. All of his personal belongings, including his correspondence
and documents that bore witness to his lifelong scientific endeavor, were lost when his
house was completely destroyed in a devastating air raid. But the most severe blow was
inflicted on him between the time he agreed to write the autobiography and the time he
completed it. His son Erwin was executed for participating in a plot to assassinate Hitler.

Despite these unusually difficult personal circumstances, Planck found the strength
to complete his scientific autobiography. He submitted the first version in March 1945,
but it first appeared only posthumously in 1947.[2] It was published together with four articles, based on lectures delivered during the last years of Planck's life: "Phantom Problems
in Science," "The Meaning and Limits of Exact Science," "The Concept of Causality in
Physics," and "Religion and Natural Science." These articles, together with the *Scientific
Autobiography*, present Planck's scientific worldview based on his philosophical and epistemological thinking. Reading these texts, we can identify with Schilpp's remark in the
introduction to the Einstein volume, regretting to the point of considering it a tragedy,
that Max Planck was too seriously ill to be able to contribute an essay to that volume.

Reading these texts, we are convinced that Planck could have been a natural subject for an additional philosopher-scientist volume in the LLP.

Einstein and Planck, at an advanced age, at about the same time, sat down to write brief accounts of their lives as scientists. They were persuaded to do it by Abderhalden and Schilpp, and they agreed to do so because they felt that this was their duty to the scientific community and the general public. Thus, thanks to the initiatives of these two academic entrepreneurs, we have today two illuminating documents of introspection and reminiscences that shed light on the intellectual development of these two great scientists and exceptional human beings. Both autobiographies focus on their scientific careers. Yet, the two autobiographies also include remarks and passages that help us to understand the shaping of their personalities and worldviews.

As a student at the Maximilian Gymnasium in Munich, Planck did well in all disciplines. In contrast to Einstein, who was very critical of his high school education, Planck praised the excellent instruction he received from his mathematical teacher in the gymnasium. He could have chosen a career in mathematics, in history, in music, or he could have followed the family line as a theologian or jurist. Instead, he devoted himself to physics. In the *Scientific Autobiography*, he explains this choice: "My original decision to devote myself to science was a direct result of the discovery which has never ceased to fill me with enthusiasm since my early youth—the comprehension of the far from obvious fact that the laws of human reasoning coincide with the laws governing the sequences of the impressions we receive from the world around us; that, therefore, pure reasoning can enable man to gain an insight into the mechanism of the latter."[3] This introductory statement summarizes Planck's attitude toward the problem of the comprehensibility of the external world.

To this, he adds a significant statement: "In this connection, it is of paramount importance that the outside world is something independent from man, something absolute, and the quest for the laws which apply to this absolute appeared to me as the most sublime scientific pursuit in life."[4] The emphasis on the independent existence of the physical world is a basic element of the world picture that emerged and crystallized in the course of an intensive, and sometimes bitter, debate between Planck and the Austrian physicist-philosopher Ernst Mach in the years 1908–1910.

The publication of Einstein's special theory of relativity immediately drew Planck's attention. He clarifies that his avowed interest in that theory does not contradict his conviction about the existence of the absolute in the physical world: "For everything that is relative presupposes the existence of something that is absolute, and is meaningful only when juxtaposed to something absolute." And he adds: "All our measurements are relative. The material that goes into our instruments varies according to its geographic source; their construction depends on the skill of the designer and toolmaker; their manipulation is contingent on the special purposes pursued by the experimenter. Our task is to find in all these factors and data, the absolute, the universally valid, the invariant that is hidden in them."[5]

We shall not compare the philosophical and epistemological conviction expressed in the two scientific autobiographies. An attempt to do that is contained in the article by Ilse Rosenthal-Schneider, "Presuppositions and Anticipations," in Schilpp's *Albert Einstein: Philosopher-Scientist* volume. The summary of Planck's scientific worldview may be formulated as follows: Planck believed in the existence of a real world; it is this real world that he presupposed as the "absolute"; and it is this absolute that he tried to establish in physics—in the absolute value of concepts, such as energy or entropy, or even of the space-time metric.

Planck concludes his *Scientific Autobiography* thus:

I have satisfied my inner need for bearing witness, as fully as possible, both to the results of my scientific labors and to my gradually crystallized attitude to general questions—such as the meaning of exact science, its relationship to religion, the connection between causality and free will—by always complying willingly with the ever increasing number of invitations to deliver lectures before Academies, Universities, learned societies, and before the general public, and these lectures have been the source of many a valuable personal stimulation which I shall gratefully cherish in loving memory for the rest of my life.[6]

Planck's attitude to the general questions, cited above, is described in the "other papers" accompanying his autobiography. The last sentence shows Planck not only as the champion of science, but also as its committed missionary. Einstein had also undertaken, throughout his life, the role of a "missionary of science," both as a duty and as a source of inspiration.[7]

By the time Einstein and Planck wrote their autobiographies, the quest for a unified worldview (to be discussed in the next chapter) was no longer on the agenda of scientists and philosophers, as it had been in the 1920s and 1930s. Einstein and Planck remained essentially alone in pursuing this goal. With their deaths, the attempt to develop a unified worldview based on scientific inquiry accompanied by epistemological reflection was marginalized. Of course, it is true that just a decade later the quest for a grand unified theory of physics and also the search for a quantum theory of gravity were again at the forefront of research. But these pursuits were the concern of an expert community of physicists and no longer served as a reference frame for a comprehensive scientific, philosophical, and cultural worldview relevant to a wider audience. Maybe it is time to reconsider this common legacy of Einstein and Planck.

NOTES

1. See Hanoch Gutfreund, "Zwei der Glänzendsten Gestirne: Max Planck und Albert Einstein, " in *Berlins wilde Energien: Porträts aus der Geschichte der Leibnizschen Wissenschaftsakademie*, ed. S. Leibried, C. Markschies, E. Osterkamp, G. Stock (Berlin: De Gruyter Akademie Forschung, 2015), 310–343.
2. The English translation was published in 1949.
3. Max Planck, *Scientific Autobiography and Other Papers* (New York: Philosophical Library, 1949), 13.
4. Ibid.
5. Ibid., 46–47.
6. Ibid., 51.
7. Jürgen Renn, "Einstein as a Missionary of Science," *Science & Education* 22 (2013): 2569–2591.

PART II

THE *AUTOBIOGRAPHICAL NOTES*

COMMENTARIES

1

THE QUEST FOR A UNIFIED WORLDVIEW

Does the product of such a modest effort [of the theoretical physicist]
deserve to be called by the proud name of a theory of the universe?
In my belief the name is justified; for the general laws on which the
structure of theoretical physics is based claim to be valid for any
natural phenomenon whatsoever. . . . The supreme task of the physicist
is to arrive at those universal elementary laws from which the cosmos
can be built up by pure deduction.

—Einstein, "Motives for Research," an address delivered on Planck's
sixtieth birthday, 26 April 1918)[1]

THE QUEST FOR A UNIFIED WORLD PICTURE WAS EXTENSIVELY DISCUSSED IN GERMANY
at the beginning of the twentieth century. Specifically, Max Planck and Ernst Mach, later
joined by Max von Laue, published essays bitterly debating the issue of the "Unity of the
Physical World Picture." In 1912, thirty-four scholars signed a manifesto (*Aufruf*) published in the leading physics journal, *Physikalische Zeitschrift*, on behalf of the Society
for Positivistic Philosophy, calling for the development of "a comprehensive *Weltanschauung*,"
and thereby advancing toward "a noncontradictory total conception [*Gesamtauf-
fassung*]."[2] Among the signatories were Ernst Mach, David Hilbert, Felix Klein, and
Sigmund Freud. This was the first public statement signed by Einstein, and he took this
challenge more seriously than the other signatories. For him, the search for a unified
worldview became a constant, lifelong commitment reflecting a deep intellectual and
psychological necessity.

In an address delivered in 1918 in celebration of Max Planck's sixtieth birthday, Einstein describes in poetic language what drives totally committed and persevering scientists in their work.[3] He asks who will remain if those who do science "out of a joyful sense
of superior intellectual power" or for "purely utilitarian purposes" would be expelled
from the temple of science by the guardian angel. What brought the others to the temple
of science is a "finely tempered nature [that] longs to escape from personal life into the
world of objective perception and thought." Among them are the theoretical physicists

whose supreme task is "to arrive at those universal elementary laws from which the cosmos can be built up by pure deduction." This demands a persevering commitment, and what it requires to undertake such work is "akin to that of the religious worshiper or the lover; the daily effort comes from no deliberate intention or program, but straight from the heart." Einstein addressed these words to Planck, but they apply even more so to himself. He had undertaken a lifelong journey in search of a simple, unified worldview and was guided by the belief that "[t]he aim of science is, on the one hand, a comprehension, as *complete* as possible, of the connection between the sense experiences in their totality, and, on the other hand, the accomplishment of this aim *by the use of a minimum of primary concepts and relations.* (Seeking, as far as possible, logical unity in the world picture, i.e., paucity in logical elements.)"[4]

The *Autobiographical Notes*, more than any other text that Einstein wrote, attest to his "escape from the personal" in search of the ultimate picture of the physical world. The forty-five pages of this text are presented as a flowing sequence of concepts, ideas, and dilemmas, not divided into chapters or sections. They are written in a telegraphic, occasionally witty and simplified style. The discussion of different topics is highly interwoven, but there is an underlying timeline. Einstein came into the world of physics as a student at the time when the mechanical worldview was losing ground to the rising electromagnetic one. As a young scientist, he contributed to clarifying all the issues and puzzles that were on the agenda of physics in those days and started two revolutions—the quantum theory and the theory of relativity. He believed that the final comprehensive theory of physical reality would be based on the field concept of general relativity. He devoted the last ten years of his life, isolated from the rest of the physics community, to a despairing effort, fluctuating between optimism and pessimism, to find such a worldview. He did not achieve this goal but believed that this would be the future task of physics. The *Autobiographical Notes* show how his life effort led him to this belief. Actually, it puts all the weight of a life's effort behind this goal, which amounted to nothing less than the creation of a comprehensive scientific worldview.

Our commentary essays are not intended to replace Einstein's text. The topics to which we devote separate chapters are highly interwoven in Einstein's narrative, sometimes beginning abruptly in the course of discussion, without even starting a new paragraph. We attempt to guide the reader through the convoluted path of the evolution of his ideas and to show how it led to his expectations toward the end of his life.

Einstein once warned us: "If you want to find out anything from the theoretical physicists about the methods they use, I advise you to stick closely to one principle: don't listen to their words, fix your attention on their deeds."[5] Reading the *Notes*, we do not have to rely only on his words. The text amply illustrates what he actually did. And where it does not, or deviates from the actual events, we complement his autobiographical reminiscences of the different chapters in his scientific life with his essays and correspondence from the time of their appearance and interpret any discrepancies that occur.

The *Autobiographical Notes* focuses on Einstein's scientific world picture, but his ideas, opinions, and actions in science and outside of science are part of a more general worldview, which he himself attempted to present in several collections of articles, letters, and addresses, which he chose and compiled. We can find a concise summary of his general worldview in a three-page essay, "The World as I See It," from 1931—a collection of statements and reflections on the nature of the human being, the meaning and purpose of life, the preferred social order, and also on science, art, and religion.[6] It can be summarized as follows.

Einstein's political ideal is democracy. He strongly believes that an autocratic system of oppression soon degenerates and that dictators, even the most benevolent ones, are always succeeded by scoundrels. This is why he passionately opposes such political systems as he sees in Russia and Italy. On the other hand, he values the German political system at that time (the Weimar Republic) for its democratic constitution and the extensive provision that it makes for the individual in case of illness and need. Einstein regards class distinctions as unjustified and, in the last resort, based on force. He abhors the military system. He is committed passionately to social justice and social responsibility. Ideals of kindness, beauty, and truth give him courage to face life cheerfully. He believes that a simple, unassuming life is physically and mentally good for everybody, and he feels contempt for luxury, possessions, and external success as the goals of human efforts. Life would seem empty without the quest for the eternally unattainable in art and science. The most beautiful experience is the mystery we sense facing art and science. It is this experience that Einstein equates with true religiosity, contrary to the notion of religion based on the concept of an anthropomorphic God who rewards and punishes his creatures. Einstein does not believe in human freedom in the philosophical sense. People act under external compulsion and also under inner deterministic necessity; thus, free will is an illusion. He quotes Schopenhauer: "A man can do what he wants, but cannot want what he wants."[7]

These are the ideas and principles that guided Einstein throughout his whole life.

NOTES

1. CPAE vol. 7, Doc. 7, p. 44.
2. For a discussion of this episode in the history of science, see "The Unified *Weltbild* as Supreme Task," in Gerald Holton, *Einstein, History, and Other Passions* (Woodbury, NY: American Institute of Physics Press, 1995), 37–39. For more on the manifesto, see Gerald Holton, "Ernst Mach and the Fortunes of Positivism in America," *Isis* 83, no. 1(1992): 27–60, 37–39.
3. Einstein, "Motives for Research," CPAE vol. 7, Doc. 7.
4. Einstein, "Physics and Reality," reprinted in *Ideas and Opinions: Based on "Mein Weltbild,"* ed. Carl Seelig (New York: Bonanza Books, 1954), 290–323, here 293.
5. Einstein, *On the Method of Theoretical Physics*, The Herbert Spencer Lecture, delivered at Oxford, June 10, 1933; reprinted in *Ideas and Opinions*, ed. Seelig, 270–276, here 270.
6. Reprinted in *Ideas and Opinions*, ed. Seelig, 8–11.
7. Cited in *Ideas and Opinions*, ed. Seelig, 8.

2

"STRIVING FOR A CONCEPTUAL GRASP OF THINGS"

In a man of my type, the turning point of the development lies in the fact that gradually the major interest disengages itself to a far-reaching degree from the momentary and the merely personal and turns toward the striving for a conceptual grasp of things.

—Einstein, *Autobiographical Notes*, p. 7 [p. 158][1]

AT THE BEGINNING OF EINSTEIN'S *AUTOBIOGRAPHICAL NOTES*, THE READER ENCOUN-ters Einstein as a living philosopher as he sits down, in a particular moment of his life, to reflect about himself, allowing the reader to look over his shoulder. The text thus makes it immediately evident without further explanation that such reflections are the essence of this life but also that the man who has dedicated his own life to the task of thinking is nevertheless ready to share this experience with his readers.

Einstein then makes this intention explicit, admitting that he needed to be convinced to write his autobiographical notes by the series editor, Paul Schilpp, but emphasizing that doing so complies with his own convictions. Designating his notes as his "obituary" marks a distance from what he writes, indicating that what he offers is a simplified and perhaps embellished picture, in any case a snapshot of a life in flux, a recurrent theme in the sequel of the text. He does not refer to his notes in autobiographical terms. Several years earlier he wrote: "Autobiographies mostly arise out of narcissism or negative feelings toward others."[2]

Einstein's *Autobiographical Notes* makes hardly any direct references to concrete biographical events. The only names he mentions are those of scientists. He does not mention places, rendering this text highly non-geographic. Therefore, the few passages in which he does report on his life beyond the development of his scientific ideas deserve all the more attention. What immediately springs to notice is that all of them are way stations on his path to an intellectual life and all of them are from his childhood days. As Einstein himself makes explicit, the watershed in his development, the development of a man of his type, was a gradual disengagement of his interest from "the momentary and

"... a conceptual
grasp of things."

the merely personal" and a turning toward "the striving for a conceptual grasp of things"
(*Notes*, p. 7 [p. 158]). But how did this disengagement come into being, or rather, how did
Einstein himself account for how his life course brought him to this turning point?

Every step of his account of past experiences is accompanied by reflections on the working of memory and thinking. Einstein is aware of the role of perspective in memory, which
is always bound to a specific moment and place, that is, the very limitation to the merely
personal he has strived to overcome in the course of his life. Yet this struggle was itself part
of a deeply personal experience that he begins to unravel with his first steps into adulthood.
These steps were marked by experiences of disillusionment. He sees their origin in a realization, early in life, of the futility of human hopes and striving, but also of the egocentrism
and the cruelty of the daily struggle for subsistence—all covered up by "hypocrisy and glittering words" (*Notes*, p. 3 [p. 157]). This limitation of human activities to a struggle for life
could not satisfy a young man who considers himself as a thinking and sentient human
being, in evident need of placing his own life within a larger and meaningful framework.

From this general description, it is of course hard to tie Einstein's realization of his
internal distance from the concerns of his fellow human beings to a particular moment
of his biography or to a specific event. But when he begins to describe his first response
to this perception of his environment, it becomes clearer what he has in mind. He writes:
"Thus I came—though the child of entirely irreligious (Jewish) parents—to a deep religiousness" (*Notes*, p. 3 [p. 157]).

The distance of which the young Einstein became aware was, it seems, that between
himself and his parents and their concerns. Einstein's father was involved in business. In
1880, the Einstein family moved from Ulm to Munich, where Einstein's father, Hermann,
together with his brother, the engineer Jakob Einstein, founded a firm producing dyna-
mos, arc and incandescent lamps, as well as telephone systems. Electrotechnology and

the electrification of urban illumination were on a rapid upswing in those days, and his family's involvement in it provided the young Albert with fascinating intellectual stimuli and challenges. In 1885, the enterprise expanded with up to two hundred employees, until competition with major firms in the ever more concentrated electrical industry caused the Einsteins to move their business activities to northern Italy, after losing a major municipal contract for the electric lighting of central Munich in 1893.

While the firm and the family moved to Pavia, Italy, in 1894, fourteen-year-old Albert first had to stay at his school in Munich, where he watched the destruction of the beautiful gardens of the family mansion to make room for ugly tenements. In December, he joined his family in Italy and eventually enrolled at the Aargau Kantonsschule in Switzerland. After only two years, in 1896, the firm established in Pavia had to be sold, and a new firm was founded in Milan, which was also liquidated after two years. Yet a further venture was undertaken, installing power plants for electrical illumination in small towns close to rivers, which were used as a power source. This attempt turned out to be moderately successful and created the expectation that young Albert might join the activities of his father.

But the young Albert had meanwhile profoundly experienced the uncertainties of this business world, shaped by harsh competition and the humiliation of financial uncertainty. He preferred to abstain from it, even after his father made an attempt, during a joint trip through their new field of activity, to introduce his son to the family business. From several contemporary documents it is clear that the young adolescent had other plans for his life. In an 1896 school essay, written in French, on his future plans, he stressed that he preferred the independence of a scientific profession, mentioning his propensity for abstraction and mathematical thinking.[3] And in letters to his friends he mocked the philistine attitude of his family and praised a solitary life in the sense of the philosopher Schopenhauer.[4]

Schopenhauer may have been always an outsider to the mainstream of philosophical discourse, particularly in the mid-twentieth century when Einstein was asked to contribute his text to The Library of Living Philosophers.[5] However, when Einstein interpreted his formative experiences with the world and its authorities, Schopenhauer was often quoted, for instance, when he commented on the hypocrisy of the world surrounding him. His writings thus became one of the intellectual resources used by the young Einstein in his effort to overcome the merely personal. His biographer and son-in-law (married to his stepdaughter), Rudolf Kayser, tells us that in the late 1920s, in Berlin: "In his little study, he works beneath the pictures of three thinkers of whom he is especially fond, Faraday, Maxwell and Schopenhauer."[6]

Clearly, however, this is already a later phase. But Einstein must have sensed this distance from the bourgeois concerns of his family even earlier, at a time when he did not yet have such intellectual resources at hand. At that earlier point, around the age of twelve, he took recourse to a religious tradition that was present in his own family only in superficial form, but strengthened first by private religious instruction that started at the age of six and then, at the age of ten, by Max Talmud, a house teacher not much older than Einstein himself who had been hired by the family in 1889, according to Jewish custom. Talmud was an orthodox Jew from Lithuania, originally employed in order to instruct little Albert in the principles of Judaism. At the beginning, this young man was too successful. To the dismay of his parents, Albert wanted them to keep a kosher home and observe other Jewish religious traditions. Religion evidently fulfilled the young Einstein's need for coping with his own identity and his relation to the world on another level or, as he writes, as "a thinking and feeling being." He found this level in what he describes as the religious paradise of his youth.

What must have made an identification with the Jewish religion—which for his parents was hardly more than a conventional tradition—all the more attractive for the young Einstein was that it offered a means of distancing himself from what he considered their bourgeois concerns, in particular in times of crisis. In the hindsight of his autobiography, religion itself takes on a deeply ambivalent character: it is the lost paradise of the youth, promising to overcome the chains of the merely personal, and at the same time, a fabrication of the education machinery of the state, which implants certain views in young people.

The religious episode was short. What Einstein apparently retained from it—and for which it provided a benchmark—was the need for a worldview that overcomes the limitations of individual and egocentric perspectives. This expectation was soon satisfied by an encounter with popular scientific readings, also introduced to Einstein by Max Talmud, who became a close friend. In this way, the secular views of his parents triumphed after all over the religious rebellion of the young adolescent. Together with Talmud, Einstein read books by Ludwig Büchner and Aaron Bernstein, which introduced him to science, not just to many of its details, but most importantly to a worldview in opposition to and critical of the religious one.

In the books by Bernstein the young Einstein learned how concepts like atoms or ether could help to uncover "mysterious" and "surprising" relationships between different areas of knowledge, which were separated by specialization into various scientific disciplines. Furthermore, several of the conceptual tools used by Bernstein and other authors to establish connections between different areas of physics and chemistry left their mark, and arguably influenced Einstein's reactions to the later readings and lectures of his student days. After publishing his first paper[7] on the phenomenon of capillarity, in which he made a comparison between molecular and gravitational forces, he wrote about the ". . . glorious feeling to perceive the unity of a complex of phenomena which appear as completely separate entities to direct sensory observation."[8]

In *Autobiographical Notes*, Einstein writes that reading popular scientific books, from which he drew the conclusion that the stories of the Bible could not be true, induced a "fanatic [orgy of] freethinking," a "mistrust of every kind of authority," and "the impression that youth is intentionally being deceived by the state" (pp. 3–5 [p. 157]). Clearly, this was a consequence not just of his readings. Rather, the popular books that he devoured became a resource for his reflections on his own adolescent experience in the midst of a bourgeois world that offered—at best—material security but no spiritual satisfaction, and especially because the young Einstein saw it as deeply imbued with hypocrisy, glittering words, and intentional lies. Indeed, Bernstein's books portrayed science as a deeply political enterprise, uniting humanity beyond traditional divisions. The author himself had been a supporter of the democratic uprising of 1848, and that rebellious spirit emanates from his books.

And the books were not Einstein's only resource. Just as crucial must have been the company and alliance with a like-minded young man, Max Talmud, who had also been a political activist and who shared his fascination with science and also the hope that science offered what both religion and politics failed to offer—a human bond beyond the merely personal, beyond class and national boundaries, even transcending the world of the living by reaching out to past generations. Against this background, Einstein's lifelong bonds with friends who shared this broad vision of science—from Max Talmud as well as the members of the Olympia Academy (see below) and Michele Besso to fellow scientists such as Max Born—take on the meaning of an almost religious alliance of believers in the humanizing mission of science.

OLYMPIA ACADEMY

In 1902, Maurice Solovine, a young student of philosophy, responded to Einstein's advertisement in the *Newspaper of the City of Bern*, offering private lessons in mathematics and physics to students and pupils. Instead of lessons on physics, they exchanged ideas about problems faced in physics, and their first encounter evolved into a lifelong friendship. They decided to read together books by prominent authors and to discuss them and were soon joined by the mathematician Conrad Habicht in this endeavor. The meetings usually took place in Einstein's apartment and lasted until late evenings and sometimes even until early morning hours. They called these meetings "Akademie Olympia" (Olympia Academy). Maurice Solovine reported later that Mileva Marić, Einstein's wife, attended the meetings and listened attentively but never took part in the discussions.[9]

Together they read, among other works:

Analysis and Sensations and *Mechanics and Its Development* by Ernst Mach
Science and Hypothesis by Henri Poincaré
Grammar of Science by Karl Pearson

Ethics by Baruch Spinoza
Treatise on Human Nature by David Hume
Logic by John Stuart Mill
What Are Numbers? by Richard Dedekind

They also read literary classics such as:

Andromaque by Jean Racine
Christmas Tales by Charles Dickens
Don Quixote by Cervantes

They also debated Einstein's work on thermodynamics and special relativity.

The "Akademie" did not last long, as Habicht left Bern in 1904 and Solovine in 1905. Einstein mentioned it often as an episode that contributed to his scientific work in years to come. In 1953, Einstein wrote in a letter to Solovine: "To the immortal Olympia academy. In your short active existence you took a childish delight in all that was clear and reasonable. Your members created you to amuse themselves at the expense of your big sisters who were older and puffed up with pride. . . . We three members, all of us at least remained steadfast."[10]

The Olympia Academy: *Left to right*, Conrad Habicht, Maurice Solovine, and Albert Einstein, around 1903. Bildarchiv der ETH-Bibliothek, Zurich.

The second personal episode Einstein mentions goes even further back into his childhood, to when he was four or five years old, when his father showed him a compass. Even at the age of sixty-seven, he still recalled the feeling of wonder that the behavior of the magnetic needle caused in him because it differed from the expectations of intuitive physics. Einstein refers to this recollection in the *Notes*, with a reservation: "I can still remember—or at least believe I can remember—that this experience made a deep and lasting impression upon me" (p. 9 [p. 159]). The apparent doubt hidden in this statement reflects Einstein's warning at the outset: "Every reminiscence is colored by one's present state, hence by a deceptive point of view" (p. 3 [p. 157]). Yet, even if in his early childhood he had not been as fully conscious of the sense of wonder as he recalls it in the *Notes*, this experience did have a lasting impact. Einstein told the story of his first encounter with a compass many times throughout his life. The childhood sense of wonder never left him. In later years, congratulating a friend, the psychiatrist Otto Juliusburger, on his eightieth birthday, he wrote: "People like us die of course like everyone else, but they do not grow old as long as they live. What I mean by that is that they stand, always as curious as children, before the great mystery into which we have been immersed."[11]

The behavior of the magnetic needle enclosed in a case did not fit the expectations of little Albert, which were shaped by motions that are caused either by touch and by force or just seem to be part of the natural constitution of the world, like the falling of heavy bodies. Aristotle and his followers built an entire world picture on these seemingly self-evident properties of bodies. A deviation of an experience from intuitive physics thus represents a conflict with a fixed world of concepts and is the essence of what Einstein defines as "wonder" in his text. He sees such clashes between experiences and conceptual frameworks as a driving force of our thinking. In this case, the unexpected behavior of the needle pointed the young Einstein to the existence of a world hidden behind the appearances, a first step into a world of thoughts that offers, however, a powerful means for understanding the behavior of tangible things—this is the reason why he recounts this episode here.

"Every reminiscence is colored by one's present state."

The story is immediately followed by the account of a second wonder that Einstein experienced at the critical age of twelve while reading a geometry book. In the Middle Ages, wonder narratives are the hallmark of saint's legends, often connected with initiations and conversions. Even in the secular form that the wonder concept here plays in the autobiography of a scientist, it still includes this dimension of an initiation into a transcendent world. In this case, the encounter with geometry as a science demonstrated to the young Einstein the possibility of attaining secure knowledge about objects of experience by pure thought, another instance of the power of thought over the real world.

He remembered, for instance, that he had managed to derive the Pythagorean theorem on his own even prior to becoming familiar with the textbook. In the *Notes*, he relates his own youthful impressions of the character of abstract thinking to the epistemology of Immanuel Kant, which is imbued by a similar fascination with the power of pure thought. In hindsight, however, Einstein characterizes this power as resting on an error, that of naively identifying the entities of geometry with real objects, without realizing the existence of a mediating assumption that was unconsciously present in geometrical arguments.

After a short digression on epistemology, Einstein returns to what he again calls his "obituary" in order to continue with an account of his adolescence, again shedding off the "merely personal." But now the account of his realization of the power of thought gradually turns into an educational history—without, however, losing its redemptory bent. He mentions his initiation into calculus, which made an impression on him that was comparable to that of geometry, as well as the popular books by Bernstein that we have discussed above and whose reading filled him with "breathless tension." Then he turns to his university education, passing over his uneven school history in Munich and later in Aarau and the quarrels over his professional future.

Albert Einstein in the Luipold-Gymnasium in Munich, 1890. Albert is standing in the first row, third from right. Note that he is the only boy with a smile on his face. Albert Einstein Archives.

The narrative of his student years is shaped by a tension between physics and mathematics. The accomplished scientist who meanwhile has pinned all his hopes on the powers of mathematics is looking back at his student years, wondering what he had missed and why he had decided to give physics rather than mathematics the primacy in his life. He explains why he chose to pursue physics as a career. Mathematics appeared to him to consist of a diversity of domains of detailed specialization, any one of which could consume a lifetime. He remembered that he realized the fundamental role of mathematics for physical insights only after years of research. He refers here to his struggle with developing the theory of general relativity and in particular to his search for the field equation. In 1912, he had written to Arnold Sommerfeld that he had gained great respect for mathematics "whose more subtle parts I considered until now, in my ignorance, as pure luxury!"[12]

In physics, on the other hand, he learned early on to "scent" which of the overwhelming and insufficiently connected experimental data would lead to fundamentals and to turn

Einstein at the Gewerbeschule in Aarau, 1896. A section from his graduating class photo. Bildarchiv der ETH-Bibliothek, Zurich.

aside from everything else (*Notes*, p. 17 [p. 161]). Einstein's assistant during the time of writing the *Notes*, Ernst Straus, recalls that Einstein often told him about his dilemma during his student years about the choice between mathematics and physics. As a student, he thought that he would never be able to decide which of the many beautiful questions in mathematics were central and which were peripheral. In physics, on the other hand, he could see what the central questions were.[13] His choice of problems to work on at the beginning of his career confirms the old Einstein's recollection of young Einstein's dilemmas and choices.

These reflections on the choice of career and the choice of problems end in another of the few personal recollections still related to Einstein's intellectual development but now set into the context of institutionalized learning that proved to be as much of a presupposition as an obstacle to this development. The coercion of examinations had a deterring effect on the student Einstein, even if he granted that the Swiss situation was less oppressive than others and that he benefited from the help of a friend whose notebooks he could use for preparing himself. These notebooks have actually survived. The friend was the mathematician Marcel Grossmann, who also helped Einstein to find a position at the patent office and later helped him with the mathematics of general relativity.

In the *Notes*, Einstein again speaks about holiness and miracle, but now it is not a miracle connected with an intellectual revelation but the miracle "that the modern methods of instruction have not yet entirely strangled the holy curiosity of inquiry; for this delicate little plant, aside from stimulation, stands mainly in need of freedom; without this it goes to wrack and ruin without fail" (*Notes*, p. 17 [p. 161]). The passage ends with a remarkable picture of a beast of prey being force-fed, a passage that interprets the opposition between the enjoyment of seeing and searching, on the one hand, and coercion and duty, on the other hand, in terms of animalistic drives and brute force. Curiosity is, for Einstein, an animal drive that cannot be repressed even with violence. Although it may be a bit far-fetched, we are nevertheless tempted to connect this picture with the popular poem "The Panther" by Rainer Maria Rilke (see box), which everybody in those days read and many knew by heart.

THE PANTHER, RAINER MARIA RILKE

His gaze against the sweeping of the bars
has grown so weary, it can hold no more.
To him, there seem to be a thousand bars
and back behind those thousand bars no world.
The soft the supple step and sturdy pace,
that in the smallest of all circles turns,
moves like a dance of strength around a core
in which a mighty will is standing stunned.
Only at times the pupil's curtain slides
up soundlessly—. An image enters then,
goes through the tensioned stillness of the limbs—
and in the heart ceases to be.

—English translation by Stanley Appelbaum

NOTES

1. Here and throughout in citations of the *Autobiographical Notes*, the first page number refers to Einstein's text in Schilpp 1979, and the one in brackets indicates the page number of the text reproduced at the end of the present volume.
2. David E. Rowe and Robert Schulmann, *Einstein on Politics: His Private Thoughts and Public Stands on Nationalism, Zionism, War, Peace, and the Bomb* (Princeton, NJ: Princeton University Press, 2007), 129.
3. Einstein, 18 September 1896, CPAE vol. 1, Doc. 22.
4. Einstein to Julia Niggli, 6 August 1899, CPAE vol. 1, Doc. 51, p. 129; Einstein to Mileva Marić, 17 December 1901, CPAE vol. 1, Doc. 128, 186.
5. On Schopenhauer, see Bart Vandenabeele, *A Companion to Schopenhauer* (Chichester: Blackwell Publishing, 2012).
6. Rudolf Kayser, *Einstein: A Biographical Portrait*, first published under the pseudonym Anton Reiser (New York: Boni, 1930), 194.
7. "Conclusions Drawn from the Phenomena of Capillarity" (1901), reprinted in CPAE vol. 2, Doc. 1.
8. Einstein to M. Grossmann, 14 April 1901, CPAE vol. 1, Doc. 100, p. 166.
9. In Solovine's introduction to Albert Einstein, *Letters to Solovine* (New York: Philosophical Library, 1987), 13.
10. Einstein to Solovine, 3 April 1953, in Einstein, *Letters to Solovine*, 143.
11. Einstein to Otto Juliusburger, 29 September 1947, AEA 38-238.
12. Einstein to Arnold Sommerfeld, 29 October 1912, CPAE vol. 5, Doc. 421, p. 324.
13. Ernst G. Straus, "Reminiscences," in *Albert Einstein: Historical and Cultural Perspectives*, ed. Gerald Holton and Yehuda Elkana (Princeton, NJ: Princeton University Press, 1982), 422.

3
"MY EPISTEMOLOGICAL CREDO"

The reciprocal relationship of epistemology and science is of noteworthy
kind. They are dependent on each other. Epistemology without contact
with science becomes an empty scheme. Science without epistemology
is—insofar as it is thinkable at all—primitive and muddled.

—Einstein's "Reply to Criticisms," in Schilpp, *Albert Einstein:
Philosopher-Scientist*, pp. 683–684

EINSTEIN DEFINED THE NATURE AND SCOPE OF HIS *AUTOBIOGRAPHICAL NOTES*, STATING
that ". . . the essential in the being of a man of my type lies precisely in *what* he thinks and
how he thinks, not in what he does or suffers" (p. 31 [p. 165]). Thus, it was natural for
him to begin the account of his intellectual life with a discussion of the process of thinking, and specifically thinking that leads to scientific understanding and discovery. He
was intrigued by the question "What, precisely, is 'thinking'"? (*Notes*, p. 7 [p. 158]) and
addressed it on many occasions in his writings and in interactions with peers. It became
part of his deep and lifelong interest in epistemology, since the early stages of his intellectual development.[1]

Einstein's answer to this question in the *Notes* is shaped by his thorough familiarity
with and involvement in philosophical discussions. This effort is also accompanied by
his keen interest in psychological investigations into thinking. As a student, he had been
fascinated by philosophers such as David Hume and Ernst Mach and the possibility of
revising established scientific concepts in the light of new empirical knowledge. Later he
came to stress the character of conceptual systems as free human creations, which cannot
be fully reduced to empirical statements and their logical connections. In the *Notes*, he
mentions Hume for having realized that essential concepts, such as the causal relation
between events, are matters of convention and cannot be deduced from anything given to
us by the senses. Hume's warning message to all those who strive for assured knowledge,
in Einstein's words, reads: "The sensory raw material, the only source of our knowledge,
through habit may lead us to belief and expectation but not to the knowledge and still less
to the understanding of law-abiding relations."[2]

". . . the essential in the
being of a man of my type."

At about the same time that he was writing the *Autobiographical Notes*, Einstein also addressed this issue of the relation between knowledge and sensory experience in his contribution to the volume of The Library of Living Philosophers devoted to Bertrand Russell. A glance at this essay helps us to understand the wider philosophical context of Einstein's brief remarks on his epistemological views in the *Notes*. There he also poses the question: ". . . what precisely is the relation between our knowledge and the raw material furnished by sense impressions?" What he specifically has in mind is: "What knowledge is pure thought able to supply independently of sense perception?"[3] Although the concepts and propositions derive their meaning from their connection with sense experiences, that connection is, according to Einstein, "purely intuitive, not itself of logical nature" (*Notes*, p. 11 [p. 160]). Only "the relations between the concepts and propositions among themselves and each other are of a logical nature . . . according to firmly laid down rules" (Notes, p. 11 [p. 159]). Einstein expresses this idea even more strongly in his remarks on Russell's theory of knowledge: "As a matter of fact, I am convinced that even much more is to be asserted: the concepts which arise in our thought and in our linguistic expressions are all—when viewed logically—the free creations of thought which cannot inductively be gained from sense-experiences"[4] The "logically unbridgeable" gap between certain concepts and sense experiences is not easily noticed because of their being so intimately combined in our everyday thinking.

This insight is the basis of Einstein's own epistemological credo: "All concepts, even those closest to experience, are from the point of view of logic freely chosen posits, just as is the concept of causality, which was the point of departure for this inquiry in the first place" (*Notes*, p. 13 [p. 160]). This is in stark contrast to the epistemology of Kant, who argued that certain concepts and assertions are part of our a priori knowledge, independent of any experience. Einstein categorically rejected this position. In his book *The Meaning of Relativity* (1921), for instance, he wrote, without explicitly mentioning Kant: "I am convinced that the philosophers have had a harmful effect upon the progress of scientific thinking in removing certain fundamental concepts from the domain of empiricism, where they are under our control, to the intangible heights of the *a priori*. . . . This is particularly true of our concepts of time and space, which physicists have been obliged by the facts to bring down from the Olympus of the *a priori* in order to adjust them and put them in a serviceable condition."[5]

Closer to philosophers of conventionalism, such as Pierre Duhem and Henri Poincaré, Einstein emphasizes in the *Notes* the freedom of choice of concepts, but even more than these philosophers he stresses the creative character of this freedom in productive thinking: "all our thinking is of this nature of a free play with concepts" (*Notes*, p. 7 [p. 158]). It is only through such a process that we can derive a meaningful coherent structure from a multitude of apparently disconnected sensory experiences.

In a similar vein, Einstein discussed the formation of concepts from sense impressions and the emergence of a "real world" from such concepts and from relations between them in his fundamental essay "Physics and Reality": "The connection of the elementary concepts of everyday thinking with complexes of sense experiences can only be comprehended intuitively and it is inadaptable to scientifically logical fixation. The totality of these connections—none of which is expressible in conceptual terms—is the only thing which differentiates the great building which is science from a logical but empty scheme of concepts."[6] Although intuitive, the certainty of the connection between the concepts and sense experiences is what "differentiates empty fantasy from scientific 'truth' " (*Notes*, p. 11 [p. 160]).

As discussed in "Physics and Reality," an example of a basic concept formed from sense impressions is that of a "bodily object" to which we attribute, in our thinking, "a real existence." This concept is not identical to all the sense impressions related to it. It is created freely in our mind, but it owes its meaning and justification to the sense impressions associated with it. Einstein asserts: "The justification of such a setting rests exclusively on the fact that, by means of such concepts and mental relations between them, we are able to orient ourselves in the labyrinth of sense impressions." This orientation in the labyrinth of sense experiences is also the essence of science, and this is what Einstein means when he says: "The whole of science is nothing more than a refinement of everyday thinking."[7]

Einstein was guided by epistemological and philosophical considerations in the formulation of new theories, in the transformation of existing bodies of knowledge, and in his continued quest for a unified theoretical framework. This effort was also accompanied by his interest in the psychology of thinking. He sympathized with the views of Max Wertheimer, one of the founders of Gestalt psychology, on the holistic character of such conceptual constructions and also of the processes by which they change. Just like Wertheimer, Einstein was interested in the "productive" character of thinking. His epistemological credo thus takes the empirical roots of knowledge for granted, but stresses the creative moment of human thinking and its psychological depths beneath the conscious. Without entering into a critical dispute or polemics with contemporary philosophers, Einstein shows himself convinced, as a result of decades of critical dispute with the founding fathers of logical empiricism, especially Moritz Schlick and Hans Reichenbach, that these depths cannot be adequately understood by a philosophy of science focusing on language, mathematics, and logic. He thus positioned himself against the mainstream of his time—the tradition of the Vienna Circle and logical empiricism.

Turning to the more psychological aspects of thinking, Einstein stresses in the *Notes* that thinking begins with the formation and manipulation of concepts generated by the mental ordering of images and memory pictures produced by sense impressions and appearing repeatedly in different contexts. He finds it natural that "thinking goes on for the most part without the use of signs (words) and beyond that to a considerable degree unconsciously" (*Notes*, p. 7 [p. 158]). Verbalization of the thought process is a secondary stage and is needed only to make thinking communicable.

Not long before writing his *Notes*, Einstein summarized, in similar terms, his psychological introspective account of the nature of thinking in a letter to the French mathematician Jacques Hadamard.[8] Hadamard was then conducting a psychological survey of mathematicians to explore their mental processes. Einstein met Hadamard when he visited Paris in 1922 to lecture on his general theory of relativity. Afterward, they corresponded for more than thirty years, mainly on issues of politics, war, and human rights. Einstein's letter is a response to a series of questions posed to him by Hadamard (see box).

March 15. 1943

In connection with the studies I am now pursuing on Psychology of Invention, I should be very grateful for answers on the following questions:

A) Is your thought more or less constantly accompanied by words or other precise signs (such as algebraic ones), or does it take place without such a help?

B) Is your thought accompanied by mental pictures of any kind (visual, auditive, muscular)? If possible describe the general character of those pictures.

C) In general does your thought require auxiliaries or guides of an analogue kind; I especially mean creative thought but similar information or ordinary thought could also be useful.

D) In ordinary thought what type do you belong to (visual, auditive, motor, etc. . . .)?

Supplementary question:

E) How do respectively full consciousness and "fringe-consciousness" participate in such processes?

Kindly direct the answers to
Professor Jacques Hadamard

Einstein's full answer to question A shows that his convictions on the nature of thinking were primarily based on introspective evidence:

The words or the language, as they are written or spoken, do not seem to play any rôle in my mechanism of thought. The psychical entities which seem to serve as elements in thought are certain signs and more or less clear images which can be "voluntarily" reproduced and combined.
There is, of course, a certain connection between those elements and relevant logical concepts. It is also clear that the desire to arrive finally at logically connected concepts is the emotional basis of this rather vague play with the above-mentioned

elements. But taken from a psychological viewpoint, this combinatory play seems to be the essential feature in productive thought—before there is any connection with logical construction in words or other kind of signs which can be communicated to others.[9]

Hadamard was particularly intrigued by Einstein's response to question E: "It seems to me that what you call full consciousness is a limit case which can never be fully accomplished. This seems to me connected with the fact called the narrowness of consciousness [*Enge des Bewusstseins*]."[10] Two chapters in Hadamard's book are in fact devoted to unconsciousness and discovery. In his *Notes*, Einstein asserts that our thinking goes on ". . . to a considerable degree unconsciously. For how, otherwise, should it happen that sometimes we "wonder" quite spontaneously about some experience?" (pp. 7–9 [p. 158]). Einstein thus offers a psychological reading of the classical topos of wonder as the starting point of philosophizing. In his interpretation, the sense of "wonder" occurs whenever one is confronted with an experience that contradicts one's established unquestioned world of concepts.

Einstein makes a short digression in the *Notes* to describe two such "wonders" that he experienced in his childhood days, which we have already mentioned in the previous chapter. Let us consider them again in the present context. One occurred at the age of four or five, when he observed the behavior of the needle of a compass, which was in no way compatible with his unconscious world of concepts. The other one occurred at the age of twelve, when he received a little book on Euclidean geometry and realized that certain unexpected assertions, like the intersection of the three altitudes of a triangle meeting at one point, could be proved with certainty. Only later did he realize that this particular "wonder," based on the impression that certain knowledge of the objects of experience can be obtained by pure thinking alone and suggesting a Kantian epistemology, actually rested on problematic assumptions.

Nevertheless, the "wonder" of being able to gain surprising insights exclusively by thinking stayed with Einstein. He was indeed the master of thought experiments, which played a crucial role for several of his scientific discoveries. Thought experiments are, however, not premised on the existence of a priori knowledge but rather work with mental constructions allowing one to explore the consequences of existing knowledge and to reflect on them from a new perspective that may lead to novel insights—even without actually carrying them out as real experiments. In the *Notes*, Einstein gives some striking examples of such thought experiments and their role in productive thinking.

The first is his description of how at the age of sixteen he imagined himself riding on a beam of light and looking at another such beam propagating parallel to him. Reflecting on the seemingly paradoxical consequences of such a situation, he felt compelled to reconsider existing theories on light propagation and eventually produced, ten years later, his special theory of relativity. The other thought experiment mentioned in the *Notes* is that of a reflecting mirror suspended in the radiation field in a closed cavity. That thought experiment convinced him in a "drastic and direct way" of the reality of light quanta. We shall discuss these thought experiments in greater detail in the relevant chapters.

In a paragraph of the handwritten manuscript of the *Notes*, which he later deleted, immediately preceding the discussion of his emblematic wonder experiences, Einstein combined his epistemological views with his psychological insights in an even more

3. "MY EPISTEMOLOGICAL CREDO"

45

Page of manuscript with crossed-out paragraph. The Morgan Library & Museum, New York.

explicit way. Taking his lead from a famous phrase by Kant, Einstein outlined a conception of the scientist as philosopher that offers, at the same time, the sketch of a psychology of discovery (translated in the box on p. 46). In this paragraph, Einstein made use of the notion of a "shock effect" to denote the result of an unusual experience that contrasts with our expectations to underpin the more traditional concept of wonder. This is quite in the sense of the philosopher and literary critic Walter Benjamin, who had described, in his writings on aesthetics, the role of the shock effect in overriding established patterns of perception. For Einstein, the shock effect serves to produce an awareness of the previously unconscious machinery of thinking, thus triggering processes of reflection about its functioning. At the same time, it is the cause of that sense of wonder about the "problematic" aspects of the external world at the outset of our conception of reality. Einstein initially intended to use the term "shock effect" in the published text as well, immediately following the deleted passage, but he crossed it out and replaced it with the common term "wonder."

TRANSLATION OF THE DELETED PARAGRAPH

Kant once wrote approximately the following sentence: "The real is not given to us but put to us [*aufgegeben*] as an assignment." This sentence offers an exceedingly accurate characterization of the boundary that a consistent positivist is denied to cross. A person for whom the contents of this sentence is constantly vivid has a philosophical attitude (in the wider and original meaning of the word "philosophical"). Though Kant's sentence applies to everybody, most people are not aware of it; because they are content with unconsciously using the concepts and thoughts through which the sensually given becomes connected for us, without becoming aware of those tools and their human origin. This is the reason why they go through life without wondering too much about what they experience. They only wonder in fact when an unusual experience contrasts with expectations provided by the unconscious machinery mentioned above. A kind of shock effect is then present through which the existence of our thinking machinery can become conscious to us; the same shock effect also leads, on the other hand, to experiencing with regard to the external world—the "recognized"—a problematical element—a "wonder." But once the gaze is sharpened for this problematic topic of thinking and objective being, then, gradually, that conception of "reality" emerges by extension and generalization which is so well characterized by the above little sentence of Kant.

The sentence attributed to Kant (see box below), contained in the deleted paragraph, is also quoted in Einstein's response to the remarks of Henry Margenau (see p. 128). He refers to this phrase as containing the truly valuable Kantian doctrine, which he came to appreciate quite late. Interpreting the meaning of this statement, Einstein writes: "There is such a thing as a conceptual construction for the grasping of the inter-personal, the authority of which lies purely in its validation. This conceptual construction refers precisely to the 'real' (by definition), and every further question concerning the 'nature of the real' appears empty."[11] This statement means that theory does not discover reality but rather shapes what is being determined as real.

THE SOURCE OF THE SENTENCE ATTRIBUTED TO KANT

The sentence that Einstein attributes to Kant was not written by Kant. It is rather a concise summary of Kant's epistemological statements about the real world. As far as we know, this formulation was suggested by the German-Jewish philosopher Jonas Cohn in *Führende Denker*

(Leipzig-Berlin: G. B. Teubner, 1921, 85). Because this formulation is frequently attributed to the Kantian tradition without specifying its source, we quote it here also in the original German:

Was bei Spinoza am Anfang stand, die einheitliche Gesetzlichkeit der ganzen Welt, die innerlich notwendige Verknüpfung aller Einzelheiten zu einem Ganzen, das steht für Kant am Ende. Von einer "Welt" dürfen wir aber im Grunde nur reden, wo alle Einzelheiten zu einem Ganzen verknüpft sind. Man erkennt so, daß dem Menschen nicht eine fertige Welt gegeben, sondern daß es seine Aufgabe ist, den gegebenen Stoff sinnlicher Empfindungen immer vollständiger in die Einheit einer Welt hineinzubauen—wir dürfen sagen: *Die Welt ist uns nicht gegeben, sondern aufgegeben* [italics added].

What constituted the beginning for Spinoza, the uniform lawfulness of the whole world, the internally necessary connection of all details to make a whole, stands for Kant at the end. But properly speaking, we can only speak of a "world" where all the details are connected to form a whole. It is thus seen that man is not given a ready-made world, but that his task is to build the given material of sensory experiences ever more completely into the unity of a world; we may say: *The world is not given to us but put to us as an assignment* [italics added].

In view of Einstein's assertion that Kant's phrase and its meaning constitute the valuable part of Kant's tradition, it is not clear why he deleted this paragraph when submitting the text for publication. In any case, this thought remained with him to be used with clear emphasis in his response to Margenau's comment.[12]

In the light of Einstein's emphasis on the nature of conceptual constructions as free creations of the human mind, their success in helping us to orient ourselves in the world must itself come as a surprise, or rather, as a "wonder." This larger sense of wonder, related to the comprehensibility of the world, is thus at the core of his epistemological credo: "The very fact that the totality of our sense experiences is such that by means of thinking (operations with concepts, and the creation and use of definite functional relations between them, and the coordination of sense experiences to these concepts) it can be put in order, this fact is one which leaves us in awe, but which we shall never understand. One may say 'the eternal mystery of the world is its comprehensibility.'"[13]

The question of the astonishing "comprehensibility" of the world came up again in an exchange of letters, in 1952, between Einstein and his lifelong friend from the days of the Olympia Academy, Maurice Solovine. The latter was concerned that Einstein's position might imply that there was a doubt about the comprehensibility of the world (see box).

We thus find in the *Autobiographical Notes* a concentrated account of Einstein's lifelong quest for understanding the thought process underlying his epistemological outlook. Although he drew on the most diverse sources, reaching from correspondence with Wertheimer and Hadamard to rethinking Kant's legacy, his epistemological credo turns out not to be an eclectic mixture of different viewpoints but his own, surprisingly coherent, philosophical edifice.

SOLOVINE WROTE TO EINSTEIN

. . . I do not understand one of your remarks: that the comprehensibility of the world is an eternal mystery. What could you mean by that? Possibly, if the world is understood in the way it is meant to be understood, namely, if our worldview is truly objective or if it is determined by our characteristic sensory and intellectual nature. This in fact is an epistemological problem that is very complex and will probably never be solved. But the comprehensibility of the world, which can take different forms, is a fact that cannot be questioned. It can only concern the determination of whether a certain form has a preference over other forms and can be considered as "correct" or "true." Since we are part of nature itself, and share manifold relations, and its phenomena are subject to uniform laws that we are compelled to know and "comprehend" for our own existence, I do not see why, especially if we also consider our quest for knowledge, what we could do to not understand it. It may be possible that this comprehensibility is of a subjective rather than objective nature. In my view, the comprehensibility of the world is an innate and irrefutable necessity (19 March 1952, AEA 21-286).

Einstein responded to this concern (30 March 1952, AEA 80-877) by pointing to the difference between a purely descriptive, lexicographic account of the world and a creative scientific theory capable of explaining nature by just starting from a few axioms fabricated by humans. He thus defies Solovine's claim that a comprehension of the world is practically inevitable, a claim that leaves entirely open how the fact that we are part of nature entails the success of such creative science:

. . . You find it strange that I consider the comprehensibility of the world (to the extent that we are authorized to speak of such) as a miracle or as an eternal mystery. Well, a priori one should expect a chaotic world that is in no way comprehensible by thought. One could (yes one should) expect the world to be subjected to law only to the extent that we order it through our intelligence. Ordering of this kind would be like the alphabetical ordering of the words of a language. By contrast, the kind of order created, for instance by Newton's theory of gravitation, is wholly different. Even if the axioms of the theory are proposed by man, the success of such a project presupposes a high degree of ordering of the objective world, which one has absolutely no right to expect a priori. So here lies the "miracle" which is being constantly reinforced as our knowledge expands.

NOTES

1. The two pioneers of research on Einstein's work, Gerald Holton and John Stachel, explored Einstein's answer to this question: Holton, "What, Precisely, Is 'Thinking'? . . . Einstein's Answer," in French 1979, 153–164; Holton, "Understanding the History of Science," in Holton 1995, 185–201; Stachel, "What, Precisely, Is 'Thinking'?," in Stachel 2005, xxxiv.
2. Albert Einstein, "Remarks on Bertrand Russell's Theory of Knowledge," in *The Philosophy of Bertrand Russell*, ed. Paul Arthur Schilpp, The Library of Living Philosophers, vol. 5 (Evanston, IL: Library of Living Philosophers, 1946), 279–291, here 285. Originally published 1944.

3. Ibid., 279.

4. Ibid., 287.

5. Albert Einstein, *The Meaning of Relativity*, 5th ed. (Princeton, NJ: Princeton University Press, 1955), 2.

6. Einstein, "Physics and Reality" (1936), reprinted in *Ideas and Opinions: Based on "Mein Weltbild,"* ed. Carl Seelig (New York: Bonanza Books, 1954), 292.

7. Ibid., 291, 290.

8. Einstein to Hadamard, 17 June 1944, AEA 12-056; reprinted in *Ideas and Opinions*, ed. Seelig, 25. See also Jacques Hadamard, *A Mathematician's Mind* (Princeton, NJ: Princeton University Press, 1945).

9. Einstein, "Physics and Reality" (1936), reprinted in *Ideas and Opinions*, ed. Seelig, 25–26.

10. Ibid., 26.

11. Einstein in Schilpp, ed., *Albert Einstein: Philosopher-Scientist*, 680.

12. See Einstein's response to Margenau (pp. 126–130). See also Thomas Ryckman, "'A Believing Rationalist': Einstein and 'the Truly Valuable' in Kant," in *The Cambridge Companion to Einstein*, ed. Michel Janssen and Christoph Lehner (Cambridge: Cambridge University Press, 2014), 377–395.

13. Einstein, "Physics and Reality" (1936), reprinted in *Ideas and Opinions*, ed. Seelig, 292.

4

THE MECHANICAL
WORLDVIEW AND ITS DEMISE

"AND NOW TO THE CRITIQUE OF MECHANICS
AS THE BASIS OF PHYSICS"

> In the beginning (if there was such a thing), God created Newton's laws
> of motion together with the necessary masses and forces. This is all;
> everything beyond this follows from the development of appropriate
> mathematical methods by means of deduction.
>
> —Einstein, *Autobiographical Notes*, p. 17 [p. 161]

MECHANICS IS ONE OF THE MOST ANCIENT FIELDS OF PHYSICAL KNOWLEDGE. THE FIRST theories of mechanics go back to Greek antiquity. In the Scientific Revolution of the early modern period mechanics played a paradigmatic role, culminating in the work of Newton, which brought together the laws of terrestrial and planetary motions under a single framework. He thus created the basis for what in the eighteenth and nineteenth centuries became an encompassing mechanical world picture, extending mechanical explanations to ever more physical phenomena. As a student, Einstein was fascinated by this extension: "What made the greatest impression upon the student, however, was not so much the technical development of mechanics or the solution of complicated problems as the achievements of mechanics in areas that apparently had nothing to do with mechanics" (*Notes*, p. 19 [p. 161]). As examples for the impressive range of mechanics he adduces the theory of sound, hydrodynamics, the theory of light, the kinetic theory of heat and, in particular, the capability to deduce the basic laws of thermodynamics from a statistical theory of classical mechanics.

The prospect to found all of physics on the basis of mechanics suggested the possibility of a scientific worldview. The fact that this worldview still had gaps and shortcomings did not necessarily contradict this hope but could rather be seen as an invitation to complete the picture. This was certainly the case for the young Einstein. In the *Notes*, he also mentions the role of mechanics as a foundation of the atomic hypothesis and thus of those parts of science related to it, not just the kinetic theory of heat, but also chemistry. However, around 1900 little was known about the actual microscopic reality of chemical atoms. As a young scientist, Einstein saw the further exploration of atomistic ideas and, in particular, a demonstration of the actual existence of atoms as a most powerful tool to pursue the completion of a scientific worldview. He applied this tool in the most diverse fields of physics, capillarity theory, the theory of solutions, the electron theory of metals, gas theory, and even the theory of light. It was this encompassing perspective that eventually brought him to recognize most clearly the fissures in the mechanical worldview.

In hindsight, Einstein recounts in the *Notes*, James Clerk Maxwell and Heinrich Hertz, protagonists of the new theory of electromagnetism, were seen as having overcome this faith in a mechanical underpinning of all of physics. But this was not the way they saw it themselves, nor did they actually abandon attempts to connect their endeavor to a mechanical foundation. As we now realize, the new conceptual foundation of electromagnetism grew out of the conceptual framework of mechanics through a gradual process of detachment, by exploring the limits of the mechanical world picture. But how could a contemporary witness recognize that this process would eventually lead to a complete break and an emancipation of the new framework from the old?

For the young Einstein, it was the reading of the historical and at the same time critical analysis of mechanics by the physicist and philosopher Ernst Mach that definitely changed this perspective. From the new perspective, the original, firm belief in a mechanical world picture to be completed now amounted to a "dogmatic faith" as Einstein writes (*Notes*, p. 19 [p. 162]). This is a famous Enlightenment motif: In the *Prolegomena to Any Future Metaphysics* (1783), the philosopher Immanuel Kant remarks: "... [S]ince the origin of metaphysics so far as we know its history, nothing has ever happened which was more decisive to its fate than the attack made upon it by David Hume" and "... the suggestion of David Hume was the very thing, which many years ago first interrupted my dogmatic slumber."[1] Einstein had been reading Kant and Hume and was fascinated by the latter's insistence on the empirical roots of all our knowledge. He may have been familiar also with the motif of Hume's impact on Kant, now applying it to the impact that Mach had upon him when he was a student.

But just as Kant had been impressed by Hume and still saw the need for a further improvement of epistemology, Einstein is quick to note that, while having been deeply influenced by Mach's epistemological position, he ultimately came to recognize it as untenable. Through his own experience as a scientist he had come to the conclusion that Mach had dramatically underestimated the "constructive-speculative" character of scientific thinking. The bone of dispute was precisely the issue at the center of young Einstein's ambitious research program aiming to complete the mechanical world picture: the reality of atoms. The *Notes* is written, of course, from the perspective of hindsight, and the nuanced philosophical reasoning it offers often has the flavor of an after-the-fact rationalization. At this point, however, we can feel the aftershock of a real intellectual clash. It must have been a provocation for the young Einstein that his philosophical hero, with whose critical assessment of mechanics as the foundation of all of physics he so much agreed, denied the existence of atoms. Mach is famously said to have responded to a believer in the existence of atoms: "Have you seen one?"[2]

As if provoked by such a question, the young Einstein feverishly searched for evidence in favor of the existence of atoms—now no longer within the framework of mechanics but across the boundaries of the emerging separate continents of contemporary physics, mechanics, thermodynamics, and electromagnetism. His reading of Mach had evidently stimulated him in more than one way, shaking the dogmatic faith in mechanics as the foundation of all of physics and identifying the reality of atoms as a key question of contemporary physics, yet now conceiving atomism in a rather more "interdisciplinary" way. This intellectual experience also changed his outlook on the question of the foundation of physics. For the Einstein who had read Mach, the search for this foundation was no longer the question of completing a world picture but rather one of exploring the limits of the existing conceptual frameworks and in particular also the limits of mechanics. That these limits are not simply fixed and given but actually rather elastic makes their exploration a highly non-trivial, sometimes tedious, and sometimes thrilling exercise. In the *Notes*, Einstein therefore goes out of his way to explain how he understands the critical analysis of a scientific theory before he comes back to the actual critique of mechanics.

He identifies two perspectives under which a scientific theory may be criticized, an external and an internal one. The external perspective refers to the agreement with empirical evidence, and in particular with experimental facts; the internal to the inner structure of a theory, in particular its "naturalness" and inner simplicity. Einstein is well aware that the first point of view is not simply a matter of rejecting a theory when it is falsified, for instance, when a disagreement between a theoretical prediction and an empirical fact is discovered. He writes: "For it is often, perhaps even always, possible to retain a general theoretical foundation by adapting it to the facts by means of artificial additional assumptions" (*Notes*, p. 23 [p. 162]). He finds it even more difficult to concisely express the second point of view, to explain what exactly he means by the "inner perfection of a theory." Ultimately, he considers it a matter of weighing different qualities against each other. He writes: "I must confess herewith that I cannot at this point, and perhaps not at all, replace these hints by more precise definitions." (*Notes*, p. 23 [p. 163])

Although he does not say so explicitly, his subsequent critique of mechanics from these two points of view makes it clear that they are mutually related. In fact, the addition of artificial assumptions to a scientific theory, triggered by the accumulation of new empirical evidence, certainly diminishes its inner perfection—whatever is precisely meant by it. What Einstein does add, however, is that the second point of view, the question of inner perfection, will play an increasingly important role when it becomes more difficult to confront a physical theory with empirical evidence, as was the case with his attempts at a unified field theory then and is the situation of fundamental physics now. He also emphasizes that he is not thinking in terms of specific domains of physics but only considers theories dealing with the totality of all physical phenomena.

After this epistemological intermezzo, Einstein returns to the critique of mechanics, beginning with the first, external point of view. That he is indeed not interested in specific domains or subdisciplines but only in encompassing scientific worldviews becomes immediately clear. His first example concerning the difficulties of mechanics to satisfactorily explain the empirical evidence refers to the "incorporation" of wave optics into the mechanical picture of the world. Rather than treating wave optics on its own merits as a specialized subdiscipline of classical physics, he focuses on the problems of explaining light waves on the basis of mechanics. Such an explanation required the introduction of a hypothetical mechanical medium: the so-called ether, with specific mechanical properties that could explain the peculiarities of these waves—for instance, their transversal

". . . ether had to lead a ghostly existence alongside the rest of matter."

character, that is, the fact that they oscillate in a plane vertical to the direction of their propagation. Such an ether must resemble a solid, incompressible body yet be permeable enough in order not to disturb the motions of the planets, for example. In other words, as Einstein pointedly formulates, the ether must lead a "ghostly existence alongside the rest of matter" (*Notes*, p. 23 [p. 163]).

The situation became worse with the development of electrodynamics, of which optics became a part when Maxwell realized that light waves are electromagnetic waves. Electrodynamics gave rise to new entities described by concepts such as that of "fields" whose existence was, as it turned out after a long process of clarification, detached from mechanical processes. In spite of many attempts to account for Maxwell's equations on the basis of mechanical models, new concepts such as the field concept gained an ever-greater autonomy with regard to the mechanical foundation of physics. In the *Notes*, Einstein carefully describes this process of the emergence and the emancipation of a new conceptual framework from an older one that turns out to be more and more incapable of accounting for the new empirical facts. This process amounted to a radical break with the mechanical foundation of physics. Yet, it happened gradually and almost unnoticed, as Einstein remarks: "[T]hus mechanics as the basis of physics was being abandoned, almost imperceptibly, because its adaptation to the facts presented itself finally as a hopeless task" (*Notes*, p. 25 [p. 163]). He concludes with the observation that the triumph of the new field of physics was, however, not complete and that its heritage was an intermediate stage characterized by dualism of concepts because the mechanical foundation was not completely superseded.

Zooming in on this dualism is a good occasion for Einstein to proceed to the second point of view of his critique of mechanics, the internal perspective, examining the inner perfection of mechanics as a foundation of physics. As we shall see, he focuses again on dualism and, more generally, on the plurality of fundamental building blocks as weaknesses of the core domain of mechanical knowledge. In his view, here as well, a field concept will ultimately help in overcoming these weaknesses. The first incongruence he discusses is that between the admissibility of all frames of reference from a purely geometrical point of view and the privileged role that classical mechanics ascribes to inertial systems as a special class of reference systems. Since this special role is not intuitively plausible, physicists introduced special assumptions to justify it, such as Newton's assumption of an absolute space, but such an assumption points to an entity outside of the objects of mechanics and thus complicates the structure of the theory. The young Einstein was impressed when he read Mach's critical-historical analysis of mechanics, pointing to this incongruence. He shared Mach's conviction that inertia should be conceived as the result of an interaction of masses. At the time, it motivated him to seek a general theory of relativity that would actually generalize the classical principle of relativity to all reference frames. After the completion of the theory, he had to realize, however, in a painful and long-winded dissociation from his Machian heuristics, which lasted until the early 1930s, that this pursuit made ultimately no sense. In the *Notes*, he laconically summarizes: "Such an attempt at a resolution does not fit into a consistent field theory, as will be immediately recognized" (p. 27 [p. 164]).

Einstein does not leave it at this sober conclusion, however, but rather reformulates once more the original plausibility that Mach's argument once had for him. He draws on an analogy that his friend and lifelong interlocutor Michele Besso had suggested to him: the situation in which people have access to only a limited experience of space, bringing them to the conclusion that the vertical direction is privileged over all other directions because it is the direction in which bodies fall down to the earth. As a matter of fact, this situation is, of course, not so imaginary after all because the distinction of the vertical is actually a hallmark of intuitive physics prior to the discovery of the spherical shape of the earth—both in history and in individual development prior to instruction. For Einstein, this argument serves to pinpoint a situation in which a theory contains an internal asymmetry, in this case between a physical preference for the vertical and the geometrical isotropy of space.

He next turns to other complications in the architecture of mechanics. As in the incongruence just discussed, they are all somehow addressed by general relativity, be it that they prepared the ground for it or that they were eventually resolved by it. These complications are hence evidently of interest to Einstein in hopes of learning from them how such inner difficulties could possibly serve to supersede an existing framework in favor of a more unified one. He first addresses three mutually related points: the duality of equation of motion and force law in classical mechanics, the peculiar nature of the force law of gravitation, and the dual nature of mass. They are related because, only in the case of gravitational interaction, the mass is the only "charge" appearing both in the field equations and in the equations of motion. As a result, in classical mechanics, motion under the influence of gravitation is independent from mass. Einstein argues that the force law of gravitation can be rewritten in such a form that it becomes related to the structure of space, a hint that gravitation should actually be conceived as a property of space rather than as a force. He finally adds that the fact that classical mechanics splits energy into kinetic and potential energy constitutes an unnatural duality, as had been observed already by Heinrich Hertz.

In summary, even independently from clashes with empirical evidence, classical mechanics harbored a number of internal tensions and incongruences that pointed, however, in a precise direction—that of replacing action at a distance by the field concept and of relating gravity to the properties of space rather than explaining it as a force, as Newton had done. Einstein does not say so explicitly, but this review of the history of classical mechanics and of the mechanical worldview clearly gave him the hope that the next step, from general relativity to a unified field theory, might also be accomplished without the explicit guidance of experimental clues. In turn, this hope evidently shaped his presentation of that history.

At the end of this critical engagement with the mechanical world picture, Einstein rhetorically addresses Newton directly: "Newton, forgive me; you found just about the only way possible in your age for a man of highest reasoning and creative power. The concepts that you created are even today still guiding our thinking in physics, although we now know that they will have to be replaced by others farther removed from the sphere of immediate experience, if we aim at a profounder understanding of relationships" (*Notes*, p. 31 [p. 165]). The criticism he had directed at Newton's legacy was evidently the unavoidable price to pay for scientific progress. That his concepts are still guiding physical thinking is not to be considered a lasting achievement but a problem to be solved. Einstein thought about his own legacy in just the same way.

NOTES

1. Paul Carus, ed., *Kant's Prolegomena to Any Future Metaphysics* (1783) (Chicago: Open Court Publishing, 1902), 3, 7.
2. Cited in Daniel Greenberger, ed., *Compendium of Quantum Physics: Concepts, Experiments, History and Philosophy* (Berlin: Springer, 2009), 615.

5

THE RISE OF THE ELECTROMAGNETIC WORLDVIEW AND THE FIELD CONCEPT

"THE TRANSITION FROM ACTION AT A DISTANCE TO FIELDS"

What eventually made the physicists abandon, after hesitating a long time, their faith in the possibility that all physics could be founded upon Newton's mechanics, was the electrodynamics of Faraday and Maxwell.

—Einstein, *Autobiographical Notes*, pp. 23–25 [p. 163]

EINSTEIN CONCLUDED THE CRITICISM OF NEWTON'S MECHANICAL WORLDVIEW AND THE attempt to make it the base of all of physics with an apology to Newton. This is followed immediately by reminding the reader that *Autobiographical Notes* is essentially an obituary. An obituary is a summary of a person's life. In the case of a person of his type, who spent all his life in search of the laws that describe physical reality, what he did, and where and when his life experiences occurred, is not important. What is important, are the thoughts that guided him in his intellectual endeavor. In this spirit, Einstein shares with the reader two autobiographical experiences from his student years—how impressed he was by classical thermodynamics and how fascinated he was with the theory of electromagnetism, formulated by James Clerk Maxwell (*Notes*, pp. 31–33 [p. 165]).

From the mid-nineteenth century, both electromagnetism (or electrodynamics) and thermodynamics developed parallel to mechanics into edifices of specialized knowledge. These edifices had increasingly separated from the original continent of mechanics and then commenced to drift freely, becoming more or less independent domains of knowledge. They do share some fundamental concepts with mechanics, such as space and time, but they also contain characteristic mental models, concepts, and theories that differ from those of mechanics.

Einstein saw in classical thermodynamics the ideal physical theory of universal content that will always be valid and was guided by it in his search for new laws and principles as, for example, on his road to the special theory of relativity (see chapter 10). Maxwell's theory was his playground on the road to his "miraculous year" of 1905. It was, therefore, important for him to emphasize the intellectual roots of his scientific endeavor before embarking on the description of how it evolved throughout his life.

In the mid-nineteenth century, Maxwell showed that electrical and magnetic phenomena were closely related, unifying them in the four fundamental equations of electromagnetism that carry his name. These equations introduced the concept of electric and magnetic fields as new physical entities that exist everywhere in space. These fields are generated by electrical charges and currents and are responsible for their behavior. They are continuous functions of the space and time coordinates. Maxwell's equations relate the spatial rates of change of electric and magnetic fields to their temporal change and to the charges and currents. Such relations between rates of change of physical quantities are represented by the mathematical language of "partial differential equations."

It turned out that changing electric fields produce changing magnetic fields, and vice versa. The result is a propagating electromagnetic wave, moving at a speed of ca. 300 million meters a second. The prediction of electromagnetic radiation and its confirmation by Hertz's experiments was a triumph of Maxwell's theory. A corollary to that triumph was the realization that light is such an electromagnetic wave. This conclusion led to the incorporation of optics into the theory of electromagnetism. Einstein recalls that this development, relating the speed of light to the electric and magnetic system of units, was for him "like a revelation" (*Notes*, p. 31 [p. 166]).

In an earlier essay on the major developments in physics ("Physics and Reality," 1936) Einstein referred to the "electric field theory of Faraday and Maxwell" as "probably the most profound transformation of the foundations of physics since Newton's times."[1] It replaces material discrete particles with continuous fields as the fundamental physical entities, thus introducing a new physical worldview alongside the declining Newtonian mechanical worldview. Michael Faraday was a British physicist whose outstanding experimental work on electromagnetic phenomena is at the basis of Maxwell's equations. Einstein mentions Faraday also in the *Notes*, in order to compare the pair Galileo–Newton of the old worldview with the pair Faraday–Maxwell of the new rising worldview. The first of each pair, guided by intuition, performed the basic experiments, and the second cast them into an exact mathematical formulation. In the *Notes*, Einstein recalls the deep impression that Maxwell's theory made on him during his student days: "The most fascinating subject at the time that I was a student was Maxwell's theory. What made this theory appear revolutionary was the transition from action at a distance to fields as the fundamental variables" (p. 31 [p. 165]).

The extension of Maxwell's theory to include optical phenomena in the theory of electromagnetism caused significant difficulties. To describe optical properties of matter and phenomena like metallic conductivity, the field was assumed to exist also in the interior of matter. Empty space (ether) and the interior of a material (dielectric) body were treated on equal footing. Matter as carrier of the field could have velocity, and that would also apply to ether. Thus, the theory inherited all the problems that had been the subject of discussion in the context of optics since the beginning of the century.

All these difficulties were removed by Hendrik Antoon Lorentz, albeit at the price of a further complication of the theory. In his modified theory of electromagnetism, ether is stationary, and its behavior is determined completely by the laws of electrodynamics. To describe the electromagnetic behavior of ordinary matter, Lorentz, in addition to the ether, introduced atomism as another invisible mechanism into his theory. Matter, according to his assumption, consists of elementary particles that are electrically charged. Properties of matter, like conductivity, optical refraction, and so on, are determined by the interaction between these particles.

According to Lorentz's theory, charges interact only through the ether by generating fields and by being subject to forces exerted by those fields. Their motion is determined by the Newtonian equations of motion. Thus, Lorentz succeeded in integrating the laws of electrodynamics and of Newtonian mechanics. There is one basic difference—the force between the particles is propagated by the field and does not act at a distance.

As much as Einstein was impressed by the Maxwell-Lorentz theory, there was one caveat. The theory combines two substantially different entities—material point-particles and continuous fields. The energy of such a system is composed of two essentially different things—the kinetic energy of the particles and the field energy. Einstein begins his 1905 paper on the nature of electromagnetic radiation with this disturbing dualism: "A profound formal difference exists between the theoretical concepts that physicists have formed about gases and other ponderable bodies, and Maxwell's theory of electromagnetic processes in so-called empty space. While we consider the state to be completely determined by the positions and velocities of an indeed very large yet finite number of atoms and electrons, we make use of continuous spatial functions to determine the electromagnetic state of a volume of space, so that a finite number of quantities cannot be considered as sufficient for the complete determination of the electromagnetic state of space."[2]

Einstein hoped that this dualism could be removed in a theory in which the point-particles and their equations of motion can be derived from the field equations. This cannot be done in Maxwell's theory. There was one significant, though unsuccessful, attempt to achieve this goal. The German physicist Gustav Mie published in 1912 an influential work on the theory of matter based on a nonlinear extension of Maxwell's equations (see chapter 13). The quest for removing the particle-field dualism was one of Einstein's incentives to pursue the goal of a theory that would treat gravitation and electromagnetism within the same framework. He believed that this can be done. We shall discuss this point in chapter 13.

NOTES

1. Einstein, "Physics and Reality" (1936), reprinted in *Ideas and Opinions: Based on "Mein Weltbild,"* ed. Carl Seelig (New York: Bonanza Books, 1954), 304–305.
2. Einstein, "On a Heuristic Point of View concerning the Production and Transformation of Light," in *Einstein's Miraculous Year: Five Papers That Changed the Face of Physics*, ed. John Stachel (Princeton, NJ: Princeton University Press), 177–198, here 177.

6
PLANCK'S BLACK-BODY RADIATION FORMULA

"BUT THE MATTER HAS A SERIOUS DRAWBACK"

All my attempts, however, to adapt the theoretical foundation of physics
to this [new type of] knowledge failed completely. It was as if the ground
had been pulled out from under one, with no firm foundation to be seen
anywhere upon which one could have built.

—Einstein, *Autobiographical Notes*, p. 43 [p. 169]

AROUND THE END OF THE NINETEENTH CENTURY, ONE OF THE PROMINENT PROBLEMS
on the agenda of classical physics was a borderline problem between thermodynamics
and electrodynamics concerning the character of the electromagnetic radiation in a cavity enclosed by totally reflecting walls. The distribution of the energy of such radiation
over the different frequencies, namely, the spectrum of the radiation, had been measured
with ever greater accuracy, also because of the technical relevance of this problem for
contemporary industry (e.g. the standardization of sources of electric light). The distribution depends only on the temperature of the cavity. The phenomenon itself is known
as "black-body radiation." A black body is the mental model of an ideal source of thermal
radiation that is assumed to absorb all incident electromagnetic radiation. This concept,
coined by Gustav Kirchhoff, became the basis of theoretical and experimental studies of
electromagnetic radiation in thermal equilibrium. Until the very end of the nineteenth
century, all attempts to formulate a theory that would adequately explain the shape of
the energy distribution failed. In the *Notes*, Einstein refers to the challenge posed by the
black-body radiation problem to classical physics as a "fundamental crisis" (*Notes*, p. 35
[p. 167]). As it turned out, classical physics could account only for the low-frequency part
of the observed energy distribution. The resolution of this problem became the starting
point of quantum theory and was built on the fundamental contributions by Max Planck.

The identification of the black-body problem as a critical challenge to classical physics,
on the other hand, was largely credited to Einstein. Einstein's struggle with this crisis led

"... no firm foundation
to be seen anywhere."

to a paper dealing with the corpuscular aspects of electromagnetic radiation published in 1905. The year 1905 deserves its name *annus mirabilis* ("miraculous year") for this and two other groundbreaking papers on Brownian motion and on special relativity, respectively. These three papers, marking Einstein's Copernican revolution, have become pillars of modern physics. They had been published in a short span of three and a half months. Einstein must have thought and worked on them simultaneously. Although they differ in their phenomenological contents and apply different mathematical tools, their origins cannot be understood in isolation from each other.

Einstein's account, in the *Notes*, of his work on the three themes of his *annus mirabilis* is also closely interwoven. It is not clear if the order of presentation is related to how his resolution of these problems actually evolved. His narrative begins with a critique of mechanics and electrodynamics, described in our previous chapters, followed by attempts to construct a foundational alternative from known experimental facts. In the *Notes*, Einstein does not mention the *annus mirabilis* papers,[1] which mark the partially successful culmination of his struggle—partially, because while these papers offer breakthrough insights, they do not constitute a new foundational alternative to classical physics. The discussion in the *Notes* is confined to his toiling along the path to this goal. We shall devote separate chapters to each of these three topics.

In 1900, following five years of work on this problem, Max Planck derived a formula that described, with great accuracy, the observed frequency distribution of black-body radiation. His path to this result was anything but straight, and it did not lead him beyond the conceptual system of classical physics. In his Nobel Prize lecture he recalls: "When I look back to the time, already twenty years ago, when the concept and magnitude of the physical quantum of action began, for the first time, to unfold from the mass of experimental facts, and again, to the long and ever tortuous path which led, finally, to its disclosure, the whole development seems to me to provide a fresh illustration of the long-since proved saying of Goethe's that man errs as long as he strives."[2]

Planck chose thermodynamics as the playground of his scientific activity and remained faithful to this choice until the end of his life. His black-body radiation law was the crown achievement of his studies. Planck's formula contains a constant h, which has been named after him and is known to be one of the fundamental constants of nature, eventually becoming a hallmark of quantum theory. An additional constant in this formula, the Boltzmann constant k, the value of which could be determined from empirical data, allowed Planck to compute correctly the size of the atom from the properties of black body radiation. This was a great success of his endeavor, which he clearly recognized and which Einstein acknowledges in the *Notes*.

Thermodynamics also attracted the attention of Einstein in the early stages of his scientific career. In the *Notes*, we find a special reference to this domain of classical physics: "It is the only physical theory of universal content concerning which I am convinced that, within the framework of the applicability of its basic concepts, it will never be overthrown" (*Notes*, p. 33 [p. 165]). Einstein's first three papers, which preceded his *annus mirabilis*, were devoted to thermodynamics. We shall come back to these papers in subsequent chapters.

Einstein explains Planck's derivation in great detail without avoiding mathematical expressions, unlike in other parts of the *Notes*. Planck used Boltzmann's formula, which connects entropy with the number of microscopic arrangements compatible with a given thermodynamic (macroscopic) state. The entropy is then proportional to the logarithm of this number. Planck applied this procedure to determine the probability of the energy distribution of a set of charged resonators of a specific frequency, representing the radiation-emitting or -absorbing cavity walls. In order to be able to compute the number of the microscopic arrangements compatible with a thermodynamic state of a given energy, he divided that energy into a large but finite number of elements of equal magnitude (energy quanta). Ultimately, the derivation of Planck's formula is based on the assumption that a material body absorbs and emits radiation in discrete packets of energy, quanta of energy, determined by Planck's constant, h.

As early as 1901 Einstein expressed doubts about Planck's derivation of his radiation formula. In a letter to Mileva Marić he wrote: "It's easy to explain what is setting me against Planck's considerations on the nature of radiation. Planck assumes that a completely definite kind of resonators (fixed period and damping) causes the conversion of energy to radiation, an assumption I cannot really warm up to. Maybe his newest theory is more general."[3]

In the *Notes*, Einstein points out that the premises from which Planck derived his formula would actually lead to a different conclusion. They imply, in fact, that the average energy of a radiation-emitting and -absorbing resonator would be independent of frequency, both at low and at high temperatures. Planck's formula (*Notes*, pp. 39–41 [p. 168]) gives a frequency-independent energy only in the limit of high temperatures. Following Planck's considerations to their logical conclusion would thus refute either statistical mechanics or Maxwell's theory of electrodynamics. Most likely both would turn out to be incorrect, except in appropriate limits. Had Planck realized that, he would have probably abandoned his great discovery, Einstein claims. This is what he implies when he writes: "But the matter has a serious drawback, which Planck fortunately overlooked at first" (*Notes*, p. 41 [p. 168]). Apparently, Einstein intended to make his criticism of Planck's radiation theory public in 1904, but his good friend Michele Besso dissuaded him from it. Years later, Besso recalled this episode in a letter to Einstein: "For my part, I was your audience in the years 1904 and '05; regarding the reception of your communications on the quantum problem I deprived you of part of your fame, but in Planck provided you in turn with a friend."[4]

Einstein recalls that all this was clear to him shortly after the appearance of Planck's work, but for a time he felt helpless: "All my attempts, however, to adapt the theoretical foundation of physics to this [new type of] knowledge failed completely. It was as if the ground had been pulled out from under one, with no firm foundation to be seen anywhere upon which one could have built" (*Notes*, p. 43 [p. 169]). Einstein's study of the foundations of statistical mechanics in the years 1902–1904, to be discussed in the following chapter, provided him with the tools to explore Planck's derivation and its consequences.

Einstein is critical of Planck's derivation of the black-body radiation formula, but he does not explain how he treated this formula in 1905; neither does he mention how he derived it in 1916, in the context of his work on the interaction between atoms and electromagnetic radiation, summarized in his two groundbreaking papers: "Emission and Absorption of Radiation in Quantum Theory" and "On the Quantum Theory of Radiation."[5] The absorption of radiation by an atom is proportional to the density of radiation. Atoms emit radiation in a spontaneous random process. Einstein assumed that this process could also be stimulated by the surrounding radiation. Applying these processes to a system of atoms immersed in a radiation field, he found a simple derivation of Planck's formula. In a letter to his friend Michele Besso, he wrote: "A brilliant idea dawned on me about radiation absorption and emission; it will interest you. An astonishingly simple derivation, I should say, the derivation of Planck's formula. A thoroughly quantized affair."[6]

These further contributions to the quantum theory of radiation provided a major confirmation of the particle (photon) nature of radiation, with these particles carrying not only energy but also momentum (impulse). In a subsequent letter, Einstein wrote to Besso: "The result . . . thus obtained is that at each elementary transfer of energy between radiation and matter, the impulse of hv/c is passed on to the molecule. It follows from this that any such elementary process is an *entirely directed process*. Thus light quanta are as good as established."[7] One could think that this had clinched for Einstein the concept of light particles and he would stop reflecting on this idea. On the contrary, he thought about it endlessly throughout his life. In one of his last letters to Besso he wrote: "[T]he whole 50 years of deliberate brooding have not brought me any closer to the answer to the question 'what are light quanta?'. Nowadays any fool thinks he knows the answer, but he is mistaken."[8]

Although the concern about light quanta was evidently of key importance to Einstein's intellectual biography, none of this is mentioned in the *Notes*. This curious fact offers another hint at the intended underlying narrative. Einstein construed his *Autobiographical Notes* as a legacy, one that could be used in order to continue the intellectual journey that he had begun and that would necessarily lead to the attempt of formulating a unified field theory. He certainly realized that there were alternative paths to which the issue of light quanta would be relevant, but giving too much play to it would detract the discussion from his chosen path.

Confronted with Planck's result, it was clear to Einstein by 1905 that a substitute for classical physics was needed, but even without being able to achieve that goal, the consequences of his considerations were monumental. He developed a new perspective on the interaction between radiation and matter and explained recently discovered phenomena such as the photo-electric effect that eventually earned him the Nobel Prize. These novel effects are also described in his *annus mirabilis* paper "On a Heuristic Viewpoint concerning the Production and Transformation of Light." However, his interest in those years was less concerned with the experimental consequences but rather with the question: "What general conclusions can be drawn from the radiation formula concerning the structure of radiation and even more generally concerning the electromagnetic foundation of physics?" (*Notes*, p. 45 [p. 169]).

The discussion of black-body radiation in the *Notes* abruptly stops here, and Einstein moves to other topics. We know his answer to the question above. The discrete aspect implied by Planck's derivation of the radiation formula is not confined to emission and absorption, but applies to the nature of the radiation field itself, which, under certain conditions, can be described as a collection of discrete quanta of energy (photons). We shall explain how Einstein reached this revolutionary conclusion in chapter 9, when we discuss how his interest in the thermodynamic fluctuations of particle motion and of the radiation field led him to a "thought experiment" with the mental model of a mirror suspended in a radiation field.

For Planck, on the other hand, the notion of an energy quantum was a mathematical device that at most explained how energy was absorbed or emitted when electromagnetic radiation interacted with matter. He did not accept that energy quanta represented a physical reality and described the nature of electromagnetic radiation itself. Had he accepted that, he would have had to abandon the familiar grounds and paved roads of classical physics. Had he done so, he would have led the revolutionary transition from classical to quantum physics. Planck's formula plays exactly that role, but Planck himself left it to others, specifically to Einstein, to announce the revolution.

In May 1905, Einstein wrote to his friend Conrad Habicht: "I promise you four papers . . . , the first of which I might send you soon. . . . This paper deals with radiation and the energy properties of light and is very revolutionary."[9] Each one of the papers mentioned in this letter represented a groundbreaking achievement. Among them was the paper introducing a modified treatment of space and time, presenting the first version of Einstein's theory of special relativity. Yet he referred only to the paper that introduced the quanta of radiation energy as truly revolutionary. In this paper, Einstein boldly emphasizes the revolutionary implications of Planck's formula. In the introduction, we read: "According to the assumption considered here, in the propagation of a light ray emitted from a point source, the energy is not distributed continuously over ever-increasing volumes of space, but consists of a finite number of energy quanta localized at points of space that move without dividing, and can be absorbed or emitted only as complete units."[10]

Einstein, at this stage in his life, was not committed to any a priori set of concepts, and, whenever his analysis justified it, he was willing to replace old conceptual frameworks with new ones. Planck, on the other hand, was reluctant to accept the consequences of his own discovery. As late as 1913, Planck still hoped to rescue classical physics from the accumulating evidence pointing at its imminent collapse: "So the theoretical physics of the present may give the impression of a venerable, yet decaying old edifice, from which one component after the other is beginning to crumble away and whose foundations are even threatening to waver. And yet nothing would be more incorrect than this idea. . . . But a closer inspection reveals . . . that certain ashlars are moved out of position only to be shifted to another more appropriate and stable place, and the actual foundations of the theory are resting in more fixed and secure position than at any previous time."[11]

In the same year, in a letter to the Prussian Academy of Sciences recommending Einstein as a member, Planck praised Einstein's outstanding contributions to physics and then inserted one reservation: "That he might sometimes have overshot the target in his speculations, as for example in his light quantum hypothesis, should not be counted against him too much. Because without taking a risk from time to time it is impossible, even in the most exact natural science, to introduce real innovations."[12] If Einstein took a risk, it paid off immensely. In his own *Scientific Autobiography*, Planck recalls this phase in his scientific career: "My futile attempts to fit the elementary quantum of action somehow into the classical theory continued for a number of years, and they cost me a great deal of effort. Many of my colleagues saw in this something bordering on a tragedy."[13]

Ironically, years later, Einstein found himself in a similar situation, and the same words were used to refer to him. When quantum mechanics had become an established theory of matter and radiation, it was generally accepted that it implied a probabilistic nature of elementary processes. But until the end of his life Einstein did not accept quantum mechanics, in its probabilistic formulation, as the final theory. One of his colleagues and friends, Max Born, wrote about him: "He has seen more clearly than anyone before him the statistical background of the laws of physics, and he was a pioneer in the struggle for conquering the wilderness of quantum phenomena. Yet later, when out of his own work a synthesis of statistical and quantum principles emerged which seemed to be acceptable to almost all physicists, he kept himself aloof and skeptical. Many of us regard this as a tragedy—for him, as he gropes his way in loneliness, and for us who miss our leader and standard-bearer."[14]

NOTES

1. These papers are reprinted and discussed in John Stachel, ed., *Einstein's Miraculous Year: Five Papers That Changed the Face of Physics* (Princeton, NJ: Princeton University Press, 2005). Originally published 1998.
2. Max Planck, "The Genesis and Present State of Development of the Quantum Theory," Nobel lecture, 2 June 1920, in Nobel Foundation, ed., *Nobel Lectures: Physics, 1901–1921* (Amsterdam: Elsevier, 1967), 407–418, here 407.
3. Einstein to Mileva Marić, 10 April 1901, CPAE vol. 1, Doc. 97.
4. Besso to Einstein, 17 January 1928, AEA 7-100.
5. Einstein, "Emission and Absorption of Radiation in Quantum Theory," *Deutsche Physikalische Gesellschaft, Verhandlungen* 18 (1916): 318–323; reprinted in CPAE vol. 6, Doc. 34. "On the Quantum Theory of Radiation," *Physikalische Gesellschaft Zürich, Mitteilungen* 18 (1916): 47–62; also in *Physikalische Zeitschrift* 18 (1917): 121–128; reprinted in CPAE vol. 6, Doc. 38.
6. Einstein to Besso, 11 August 1916, CPAE vol. 8, Doc. 250.
7. Einstein to Besso, 6 September 1916, CPAE vol. 8, Doc. 254.
8. Einstein to Besso, 12 December 1951, AEA 7-401.
9. Einstein to Conrad Habicht, 18 or 25 May 1905, CPAE vol. 5, Doc. 27.
10. Stachel, ed., *Einstein's Miraculous Year*, 178.
11. Max Planck, "Neue Bahnen der Physikalischen Erkenntnis" (1913), in Max Planck, *Wege zur Physikalischen Erkenntnis: Reden und Vorträge* (Leipzig: Hirzel, 1934), 35.
12. Max Planck, "Proposal for Einstein's Membership in the Prussian Academy of Sciences," 12 June 1913, in CPAE vol. 5, Doc. 445.
13. Max Planck, *Scientific Autobiography and Other Papers* (New York: Philosophical Library, 1949), 44–45.
14. Max Born, "Einstein's Statistical Theories," in Schilpp, ed., *Albert Einstein: Philosopher-Scientist*, 163–164.

7

EINSTEIN'S STATISTICAL MECHANICS

CLOSING THE "GAP"

Not acquainted with the investigations of Boltzmann and Gibbs,
which had appeared earlier and actually exhausted the subject, I
developed the statistical mechanics and molecular-kinetic theory of
thermodynamics based upon it.

—Einstein, *Autobiographical Notes*, p. 45 [p. 170]

SEVERAL QUESTIONS RELATED TO THIS QUOTATION FROM THE *NOTES* DESERVE ATTEN-
tion: What did Einstein know about the work of Boltzmann and Gibbs when he developed his version of statistical mechanics in 1902–1904? Did their work really exhaust the
subject? What was new in Einstein's formulation of statistical physics?

Statistical mechanics applies the theory of probability to the study of physical systems
composed of a large number of microscopic constituents, specifically, of material particles. It provides a framework for relating the random motion of billions and billions of
individual atoms and molecules and their collisions to the thermodynamic properties of
macroscopic systems like temperature, pressure, and entropy. It is also a framework for
deriving the basic laws of thermodynamics.

The building blocks of statistical mechanics can be found in the numerous publications of James Clerk Maxwell and Ludwig Boltzmann, nineteenth-century pioneers of
the kinetic theory of gases (see box). It was, however, the book published in 1902 by
Josiah Willard Gibbs, *Elementary Principles of Statistical Mechanics*, that gave the first
formulation of statistical mechanics as a complete and autonomous theory. In the same
year, and in two consecutive years, Einstein published three papers on statistical mechanics: "Kinetic Theory of the Thermal Equilibrium and of the Second Law of Thermodynamics" (1902), "A Theory of the Foundations of Thermodynamics" (1903), and "On
the General Molecular Theory of Heat" (1904).[1] These papers originated independently
and almost simultaneously with Gibbs's foundational treatise. However, unlike Gibbs,

Einstein immediately explored the connections with a wide array of other topics he was pursuing at that time.

THE KINETIC THEORY OF GASES

The kinetic theory describes a gas (specifically, an "ideal gas" of sufficiently low density) as a collection of a large number of particles (atoms or molecules) that move at high velocities and collide randomly with each other and with the walls of the vessel in which they are contained. The macroscopic properties of such a system in equilibrium can be described by statistical methods. In particular, the kinetic theory of gases provides a relation between the pressure, temperature, and volume of an ideal gas. The pressure is caused by the collisions of the gas particles with the walls of the container, and the temperature of the system is related to their average kinetic energy. The kinetic theory can also explain other macroscopic features, like the viscosity and thermal conductivity of the system.

Einstein's work on statistical mechanics was guided by his conviction that atoms and molecules really exist. Although the kinetic theory of gases, based on an atomistic picture, was remarkably successful by the end of the nineteenth century, many physicists of that time did not accept this notion. One of the leading critics of the atomistic hypothesis was the Austrian physicist Ernst Mach. For him, the atomistic picture underlying the kinetic theory of gases was just a hypothesis that helped to explain experimental results. Mach was an empiricist for whom any statement that is not supported by direct observations should not be accepted as part of the description of physical reality. Einstein was strongly influenced by Mach in his formulation of the theory of relativity, the special and the general, but he did not subscribe to his anti-atomistic views. In the *Notes*, he mentioned his belief in atomism as a major incentive for his work on statistical mechanics: "My principal aim in this was to find facts that would guarantee as much as possible the existence of atoms of definite finite size" (*Notes*, p. 47 [p. 170]).

The growing significance of atomism at the turn of the century and of the conceptual problems that came with it attracted the attention of young Einstein. For him, atomism was a bond between different fields of contemporary science that allowed him to cherish the hope for a conceptual unification of different phenomena. He was firmly convinced of Boltzmann's atomistic principles. In a letter to Mileva Marić from 1900 he wrote: "The Boltzmann is magnificent. . . . I am firmly convinced that the principles of the theory are right, which means that I am convinced that in the case of gases we are really dealing with discrete mass points of definite finite size."[2]

Another innovative aspect of Einstein's formulation of statistical mechanics was his understanding that fluctuations of physical quantities should be taken seriously; they are important and may lead to new consequences. The thermodynamic properties of a state of a physical system are derived as statistical averages over all the possible microscopic configurations of the positions and velocities of the particles composing the system. Boltzmann and Gibbs argued that fluctuations around these averages are exceedingly small and claimed that they would never be observed in a macroscopic system.

Boltzmann asserted: "Even in the smallest neighborhood of the tiniest particles suspended in a gas, the number of molecules is already so large that it is futile to hope for any observable deviation, even in a very small time, from the limits that the phenomena would approach in the case of an infinite number of molecules."[3] Einstein did not accept this conclusion and was looking for cases where such fluctuation phenomena, predicted by statistical mechanics, could be observed; otherwise, there would be no need for statistical mechanics.

Einstein studied Boltzmann's book *Lectures on Gas Theory* but was not aware of Boltzmann's papers in which his work is motivated and further elaborated. He read the book with his own interests in mind and developed a novel interpretation of its results by placing them in a new and broader context including the theory of radiation and the theory of electrons in metals, topics to which we shall return below. In a letter to Mileva Marić he wrote: "At present I am again studying Boltzmann's theory of gases. Everything is very nice, but there is too little stress on the comparison with reality."[4] A few months later, he wrote to his friend Marcel Grossmann: "Lately I have been engrossed in Boltzmann's work on the kinetic theory of gases and these last few days I wrote a short paper myself that provides the keystone in the chain of proofs that he had started."[5] The "short paper" is probably an earlier version of Einstein's first paper in his statistical mechanics trilogy.

These two comments indicate that Einstein detected certain shortcomings in Boltzmann's work, although in the discussion of the achievements of classical mechanics, he states: ". . . it was also of profound interest that the statistical theory of classical mechanics was able to deduce the basic laws of thermodynamics, something in essence already accomplished by Boltzmann" (*Notes*, p. 19 [p. 162]).

Einstein is more explicit about the shortcomings in the previous work on statistical mechanics in the opening remarks of his 1902 paper. It begins with the statement that in spite of the great achievements of the kinetic theory of heat, "the science of mechanics . . . has not yet succeeded in deriving the laws of thermal equilibrium and the second law of thermodynamics using only the equations of mechanics and the probability calculus, though Maxwell's and Boltzmann's theories came close to this goal. The purpose of the following considerations is to close this gap."[6] Einstein mentions Maxwell and Boltzmann, but he does not mention Gibbs. Einstein was ignorant of the fact that his paper contained the essential features of statistical mechanics that had been thoroughly discussed by Gibbs a year earlier.

The underlying concept applied in Einstein's three papers on statistical mechanics is related to the method of averaging over the fast, random changes of the microscopic state of a system of N particles. Such a state is characterized at every moment in time by the positions and velocities of all the particles. Einstein (and Gibbs) argued that instead of averaging over the temporal changes of this immense set of parameters, one can assume an imaginary "ensemble" of all possible microstates of the system and perform the average over them. Einstein discussed such ensembles of microstates of constant energy and showed that temporal and "ensemble" average give the same results.

In the first paper, Einstein derived the second law of thermodynamics from the laws of mechanics and the theory of probability. He explored the statistical-mechanical description of temperature and entropy and derived the equipartition theorem (see box), which asserts that, in equilibrium, the energy of the system is equally distributed between its microscopic degrees of freedom. Einstein concluded that the mechanical aspects of the system do not play a significant role, indicating that the results may be more general. His second paper essentially frees his statistical mechanics from mechanics, paving the way for many applications to systems outside of mechanics, such as the radiation field, electrons in metals, and others.[7]

THE EQUIPARTITION THEOREM

The equipartition theorem is one of the basic results of classical statistical mechanics. It shows that the energy of a dynamical system in equilibrium is equally distributed between its degrees of freedom and determine its energy. Degrees of freedom are parameters that describe the microscopic state of the system. In an ideal gas of pointlike atomic particles, they are the $3N$ components of the velocities of all the N particles. If the particles are composite molecules, then there are also $3N$ degrees of freedom of the rotational motion of the molecules. If the particles are also subject to an external force, which depends on their position, then the spatial coordinates of the particles are also counted in the number of the system's degrees of freedom. The energy associated with each degree of freedom is $(\frac{1}{2})kT$, where k is Boltzmann's constant and T is the temperature.

Having completed the second paper, Einstein wrote to his friend Michele Besso: "After many revisions and corrections, I finally sent off my paper. But now the paper is perfectly clear and simple, so that I am quite satisfied with it. The concepts of temperature and entropy follow from the assumption of the energy principle and the atomistic theory, and so does the second law in its most general form, namely the impossibility of a *perpetuum mobile* of the second kind, if one uses the hypothesis that state distributions of iso. [isolated] systems never evolve into more improbable ones."[8]

There is significant overlap between Einstein's three papers on statistical mechanics, at least as far as the general method is concerned. However, the third paper contains important new results. Einstein derives there the fluctuations in the energy of a system in contact with another system of a very large energy and constant temperature T. He shows that the average value of the energy fluctuations is related to the constant k, known as the Boltzmann constant, which plays a central role in the kinetic theory of heat and has already been discussed in the previous chapter. This relation provides a new meaning to the constant k. Einstein then applies this calculation to the energy of the black-body radiation field enclosed in a cube of side length L, and shows that if L is equal to the wavelength, λ_m, of the maximum in the energy distribution of the radiation field, then energy fluctuations are of the order of magnitude of the energy. This result leads to the surprising conclusion: "One can see that both the kind of dependence on the temperature and the order of magnitude of λ_m can be correctly determined from the general molecular theory of heat, and considering the broad generality of our assumptions, I believe that this agreement must not be ascribed to chance."[9]

Einstein's formulation of statistical mechanics became a cornerstone of his work toward the "miraculous year." As we shall discuss below, it provided the basis for his revision of the foundations of classical physics and his analysis of black-body radiation, of Brownian motion, and other fluctuation phenomena, as evidence for the reality of atoms.

There is another topic that attracted Einstein's interest in the framework of his atomistic studies in those years, but is not mentioned in the *Notes*. In analogy to the freely moving atoms of a gas, the German physicist Paul Drude developed a theory of electrons in metals. Drude's theory assumes freely moving charge carriers in a metal, accounting both for its electric and its thermal conductivity, as well as for the connection between these conductivities, described by the so-called Wiedemann-Franz law. Einstein was critical of Drude's theory, but we do not know the exact nature of his objections, as we do not know the contents of a letter Einstein wrote to Drude in 1901 and we also do not know the contents of Drude's response. All we know about these letters and about Einstein's interest in this subject is based on the love letters exchanged with Mileva Marić. Einstein might have developed, independently, his own version of the electron theory of metals. His interest in the electron theory played, in any case, a role in the emergence of his formulation of

Einstein with his first wife, Mileva Marić, in Kac, Serbia (1910), where Mileva's father owned a country house. Bildarchiv der ETH-Bibliothek, Zurich.

statistical mechanics because it pointed to the necessity to derive the equipartition theorem for a larger class of physical systems, beyond the kinetic theory of gases.[10]

NOTES

1. Reprinted in CPAE vol. 2, Docs. 3, 4, 5.
2. Einstein to Marić, 13[?] September 1900, CPAE vol. 1, Doc. 75.
3. Quoted in Martin Klein, "Fluctuations and Statistical Physics in Einstein's Early Work," in *Albert Einstein: Historical and Cultural Perspectives*, ed. Gerald Holton and Yehuda Elkana (Princeton, NJ: Princeton University Press, 1982), 44.
4. Einstein to Marić, 30 April 1901, CPAE vol. 1, Doc. 102.
5. Einstein to Grossman, 6(?) September 1901, CPAE vol. 1, Doc. 122.
6. Albert Einstein, "Kinetic Theory of Thermal Equilibrium and of the Second Law of Thermodynamics," CPAE vol. 2, Doc. 3.
7. See the discussion of this point in Klein, "Fluctuations and Statistical Physics in Einstein's Early Work," 39.
8. Einstein to Besso, 22(?) January 1903, CPAE vol. 5, Doc. 5.
9. Albert Einstein, "General Molecular Theory of Heat," *Annalen der Physik* 14 (1904): 354–362; reprinted in CPAE vol. 2, Doc. 5, p. 77.
10. For a discussion of these issues, see Jürgen Renn, "Einstein's Controversy with Drude and the Origin of Statistical Mechanics," *Archive for History of Exact Sciences* 51, no. 4 (December 1997): 315–354.

8
BROWNIAN MOTION

"THE EXISTENCE OF ATOMS OF DEFINITE FINITE SIZE"

> In the midst of this I discovered that, according to atomistic theory,
> there would have to be a movement of suspended microscopic particles
> capable of being observed, without knowing that observations
> concerning the Brownian motion were already long familiar.
>
> —Einstein, *Autobiographical Notes*, p. 47 [p. 170]

EINSTEIN CONCLUDES HIS DISCUSSION OF PLANCK'S RADIATION FORMULA BY POSING A question about the consequences that can be drawn from this formula concerning the electromagnetic foundation of physics. Without answering this question, he makes an abrupt transition, not even starting a new paragraph, to the issue of Brownian motion (see box).

". . . physiologists have observed (unexplained) motions of small, inanimate bodies."

BROWNIAN MOTION

Brownian motion is the irregular motion of a microscopic particle suspended in a liquid. The first systematic investigation of such motion can be traced back to the botanist Robert Brown, who published his careful observations in 1828. He investigated a large number of different particles suspended in a fluid—from plant pollen to fragments of an Egyptian sphinx—and explored also a multitude of possible causes—from currents in the fluid to interaction among the particles to the formation of small air bubbles. In this way, Brown and his successors succeeded in excluding many potential explanations of the irregular motion of the suspended particles, in particular, the idea that they may be an exclusive property of organic matter and in some way an expression of "life." Still, Brownian motion did not become a topic of wide interest among physicists, at least not until the middle of the nineteenth century. A series of articles appeared in the meantime about the influence of special circumstances on Brownian motion, for example, the temperature of the liquid, capillarity, convection currents in the fluid, evaporation, light incident on the particle, electric forces, or the role of the environment. Since the middle of the nineteenth century, scientists considered the kinetic theory of heat, which had become an increasingly useful tool for explaining thermal phenomena on a mechanical basis, as a possible explanation of Brownian motion. It was plausible to assume that the erratic motion of the suspended particle was caused by impacts of the randomly moving molecules of the host liquid. However, none of these efforts led to a consistent theory of Brownian motion. As it turned out, everybody who attempted to cope with its explanation did not realize that Brownian motion does not have a velocity in the classical sense, because it is a stochastic process in which the mean-square displacement is proportional to time. An exception was the Polish physicist Marian von Smoluchowski, who developed an adequate theory of Brownian motion at about the same time as Einstein.

We have already quoted Einstein's statement in the *Notes* that his major goal was to establish the existence of atoms of definite size. His attempts to achieve this goal, combined with the quest of observing fluctuations of a physical system such as thermal radiation, motivated by his interest in Planck's groundbreaking work, constituted a driving force for his work leading to the "miraculous year" of 1905. Initially, he thought that the scenario of electromagnetic radiation enclosed in a cavity offered the best opportunity to discover such fluctuations (we shall discuss this issue again in chapter 9, when we get back to Einstein's conclusion about the corpuscular nature of the radiation field). However, his search for observable fluctuations led him to a completely different system. He concluded that colloidal particles, large enough to be observed under a microscope, suspended in a liquid, undergo a perpetual random motion owing to the thermal motion of the molecules. The latter motion could be described by the molecular-kinetic theory.

In the *Notes*, Einstein recalls the key step that led him to this conclusion. He conjectured that particles, suspended in a liquid, contribute to the osmotic pressure in the same way as do molecules—for example, molecules of salt or sugar—dissolved in the liquid. Osmotic pressure in classical physics is the pressure exerted on a membrane immersed in a fluid, which contains on one side a substance in solution that cannot pass through the membrane. This concept was originally introduced only for the

thermodynamics of solutions to deal with dissolved molecules, and it was well understood at that time. However, its applicability to a collection of suspended particles was
not an obvious step.

Having taken this step, Einstein could apply the concepts and methods with which he
was familiar from his earlier work. His first two papers, published before the three papers
on statistical mechanics, to which Einstein referred later as "my worthless beginnings,"
were not so worthless, after all. In particular, the second of these papers, with the long
title "On the Thermodynamic Theory of the Difference in Potentials between Metals and
Fully Dissociated Solutions of Their Salts and on an Electrical Method for Investigating
Molecular Forces,"[1] already dealt with several of the topics that would play a key role in
his paper on Brownian motion. For example, it dealt with the nature of diffusion and
with the application of thermodynamics to the theory of solutions. In 1903, Einstein had
already developed an idea of how to calculate the size of ions in a liquid by using hydro-
dynamic arguments and how to determine the size of neutral salt molecules by using diffusion. This idea developed into his doctoral thesis, "A New Determination of Molecular
Dimensions,"[2] which he concluded successfully in 1905 and submitted to the University
of Zurich, after earlier failed attempts.

In his doctoral dissertation, Einstein arrived at an expression for the diffusion coefficient that contains the size of an atom. He used this equation together with a hydrody-
namic equation that relates atomic size and the viscosity of the liquid, to derive values
for the size of atoms from experimental data about diffusion and viscosity. In his paper,
which led to the description of Brownian motion, Einstein derived again the viscosity-
diffusion equation, this time with the methods of statistical physics, because only in this
way could he justify the application of concepts like osmotic pressure to a collection of
suspended particles. He combined the results of his thesis research with those he collected in the study of fluctuations in the context of statistical mechanics. Thus, he had all
the connections at hand to build a model of observable fluctuations in a material system,
as exemplified by Brownian motion.

The result of this accumulated effort was Einstein's derivation of the mean square distance covered by such particles in a given time, which he published in one of the *annus
mirabilis* papers, "On the Motion of Small Particles Suspended in a Stationary Liquid
Required by the Molecular-Kinetic Theory of Heat."[3] Experimental measurement of the
predicted displacement would provide a new and reliable method for determining the
Avogadro number (see box) and the actual size of atoms.

In the introductory remarks to this paper, Einstein writes: "It is possible that the move-
ments to be discussed here are identical with the so-called Brownian molecular motion,
however the information available to me regarding the latter is so lacking in precision
that I can form no judgment in the matter." The science historian Martin Klein, referring
to this paper, remarks: "Einstein had *invented* the Brownian motion. To say anything less,
to describe this paper in the usual way, that is, as his *explanation* of the Brownian motion,
is to undervalue it."[4]

In the letter to Conrad Habicht mentioned in chapter 6, Einstein writes about this
paper: "The third [paper] proves that, on the assumption of the molecular theory of heat,
bodies on the order of magnitude 1/1000 mm, suspended in liquids, must already perform an observable random motion that is produced by thermal motion; in fact, physiologists have observed (unexplained) motions of suspended small, inanimate, bodies,
which motions they designate as 'Brownian molecular motion.'"[5]

THE AVOGADRO NUMBER

The Avogadro number or the Avogadro constant is the number of particles, usually atoms or molecules, in one mole of a substance. The mole is defined as the amount of substance, which contains as many constituent particles, atoms, or molecules, as there are in 12 grams of the common carbon isotope ^{12}C. The Avogadro number is named after the Italian scientist Amedeo Avogadro (1776–1856), who was the first to suggest that the volume of a gas of any material, at a given temperature and pressure, is proportional to the number of atoms or molecules in that volume, but he did know the proportionality constant's value. The latter is given by the Avogadro number. The French physicist Jean Perrin won the 1926 Nobel Prize in Physics largely for measuring the Avogadro number by several different methods. Its value is 6.022×10^{23}. Einstein recognized the far-reaching implications of his paper on Brownian motion right from the beginning.[6] In the introduction, he concludes that if the predicted behavior of particles suspended in a liquid can be observed, then: ". . . classical thermodynamics can no longer be viewed as applying to regions that can be distinguished even with a microscope, and an exact determination of actual atomic sizes becomes possible. On the other hand, if the prediction of this motion were to be proved wrong, this fact would provide a far-reaching argument against the molecular-kinetic conception of heat."[7]

In a series of ingenious experiments, first published in 1908, Jean Perrin confirmed experimentally the predictions of this paper. Thus, Einstein achieved his goal of finding "facts which would guarantee as much as possible the existence of atoms of definite finite size." In the *Notes*, Einstein emphasizes the importance of this result to convince the anti-atomists, like Wilhelm Ostwald and Ernst Mach, of the reality of atoms. We have already mentioned the anti-atomistic views of Mach. Ostwald was another leading proponent of this opinion and a prominent supporter of a competing scientific worldview centered on the concept of energy and hence called "energetics." Around the turn of the century, the discussion of energetics led to a violent controversy between Ostwald and the main proponent of atomism, Ludwig Boltzmann.

Einstein attributes the antipathy of these scholars toward atomic theory to their positivistic philosophical attitude and asserts that even scholars of such standing "can be hindered in the interpretation of facts by philosophical prejudices." He concludes this assessment with a remark consistent with his epistemological credo, discussed in chapter 3: "The prejudice—which has by no means disappeared—consists in the belief that facts by themselves can and should yield scientific knowledge without free conceptual construction. Such a misconception is possible only because one does not easily become aware of the free choice of such concepts, which, through success and long usage, appear to be immediately connected with the empirical material" (*Notes*, p. 47 [p. 170]).

In his remarks on Bertrand Russell, Einstein refers to this difficulty as a "plebeian illusion": "This more aristocratic illusion concerning the unlimited penetrative power of thought has as its counterpart the more plebeian illusion of naïve realism, according to which all things 'are' as they are perceived by us through our senses. This illusion dominates the daily life of men and of animals; it is also the point of departure in all of the sciences, especially of natural sciences. The effort to overcome these two illusions is not independent the one of the other."[8] By "aristocratic illusion," Einstein refers to the philosophical belief in the unlimited power of thought, such that everything that can be known can be found by pure reflection.

Arnold Sommerfeld recalled that the "old fighter against atomistics," Wilhelm Ostwald, told him once that he had been converted to atomistics by the complete explanation of Brownian motion.[9] In the same volume, Max Born remarked: "I think that these investigations of Einstein have done more than any other work to convince physicists of the reality of atoms and molecules, of the kinetic theory of heat, and of the fundamental part of probability in the natural laws. Reading these papers one is inclined to believe that at that time the statistical aspect of physics was preponderant in Einstein's mind; yet at the same time he worked on relativity where rigorous causality reigns."[10]

NOTES

1. Reprinted in CPAE vol. 2, Doc. 2.
2. Reprinted in CPAE vol. 2, Doc. 15.
3. Reprinted and discussed in John Stachel, ed., *Einstein's Miraculous Year: Five Papers That Changed the Face of Physics* (Princeton, NJ: Princeton University Press, 2005), 71–85. Originally published 1998.
4. Martin Klein, "Fluctuations and Statistical Physics in Einstein's Early Work," in *Albert Einstein: Historical and Cultural Perspectives*, ed. Gerald Holton and Yehuda Elkana (Princeton, NJ: Princeton University Press, 1982), 47.
5. Einstein to Habicht, 18 or 25 May 1905, CPAE vol. 5, Doc. 27.
6. See Mary Jo Nye, *Molecular Reality: A Perspective on the Scientific Work of Jean Perrin* (London: MacDonald, 1972).
7. Stachel, ed., *Einstein's Miraculous Year*, 85–86.
8. Einstein in Schilpp, ed., *The Philosophy of Bertrand Russell*, 281.
9. Arnold Sommerfeld, "To Albert Einstein's Seventieth Birthday," in Schilpp, ed., *Albert Einstein: Philosopher-Scientist*, 105.
10. Born in Schilpp, ed., *Albert Einstein: Philosopher-Scientist*, 166.

9

A REFLECTING MIRROR IN RADIATION FIELD

"THE MIRROR MUST EXPERIENCE CERTAIN RANDOM FLUCTUATIONS"

> This way of looking at the problem showed in a drastic and direct way
> that a type of immediate reality has to be ascribed to Planck's quanta,
> that radiation must, therefore, possess a kind of molecular structure as
> far as its energy is concerned.
>
> —Einstein, *Autobiographical Notes*, p. 51 [p. 171]

THE WORK ON BROWNIAN MOTION, LEADING TO THE DERIVATION OF THE AVERAGE DIS-placement of a relatively large particle suspended in a liquid of randomly moving mole-cules, suggested to Einstein an analogy with an apparently very different physical system. In one of his classical thought experiments, he imagined a reflecting mirror immersed in the radiation field enclosed in a cavity. This analogy allowed him to resume the discussion of Planck's formula of the spectral energy distribution of black-body radiation.

According to Maxwell's classical theory, an electromagnetic wave carries energy and momentum. The change in momentum of an electromagnetic wave caused by reflection from a material surface generates a force on that surface. The radiation field in a cavity is a superposition of infinitely many such wavelets interfering with each other and moving in all directions. Their interaction with a reflecting surface generates a radiation pressure on that surface. In his 1904 paper "The General Molecular Theory of Heat," Einstein derived the energy fluctuations of such a radiation field. By a similar calculation he could also derive the fluctuations in the radiation pressure caused by the momentum fluctuations.

Now the analogy with the Brownian motion problem becomes apparent. In Einstein's thought experiment, the suspended mirror is free to move in the direction perpendicular to its surface. If the mirror were for some reason in motion, and if there were no pressure fluctuations, it would gradually slow down, because the reflection on the front side would cause a stronger force than on the back side. This force imbalance is analogous to

"The mirror . . . must experience certain random fluctuations."

the slowing-down force acting on a particle suspended in a liquid caused by the liquid's viscosity. However, the pressure fluctuations would not allow it to come to rest. This was Einstein's crucial insight, which led him to the conclusion that fluctuations of heat radiation can be related directly to the material motion of a mirror suspended in a cavity filled with radiation. As a consequence of the incident radiation and the friction force resulting from radiation pressure, the mirror should exhibit a behavior similar to Brownian motion.

According to the equipartition theorem (see chapter 7), the average kinetic energy assigned to every degree of freedom of motion of all the constituents of a given system is equal to $(\frac{1}{2})kT$. Thus, this is the average kinetic energy of the mirror, free to move in one direction only (a single degree of freedom). Einstein could show that the radiation pressure variations calculated in the framework of Maxwell's theory, based on the notion of a continuous electromagnetic field, are not sufficient to set the mirror in motion with this amount of energy. On the other hand, if one acknowledges that there exists a second type of pressure fluctuations, which cannot be derived from Maxwell's theory, being caused by the corpuscular nature of the radiation, then the expected motion of the mirror follows naturally. This result was for Einstein a "drastic and direct way" to show that Planck's quanta are real.

This thought experiment is not mentioned in Einstein's 1905 papers, nor in his correspondence with Mileva Marić. It was discussed in detail, for the first time in print, in 1909.[1] However, in later recollections Einstein pointed out that he already had this idea in the early 1900s. This he claimed in the *Notes* and a few year later, in a letter to Max von Laue: ". . . in 1905 I was already sure that [Maxwell's theory] led to erroneous fluctuations in radiation pressure and thus to an incorrect Brownian motion of a mirror in Planck's blackbody radiation. In my opinion, one cannot avoid attributing to radiation an objective atomistic structure, which naturally does not fit into the frame of Maxwell's theory."[2]

Einstein argued that his analysis of the fluctuating mirror in a radiation field provided convincing evidence for the existence of light quanta. However, Max Planck was not convinced. In a discussion that followed Einstein's 1909 lecture on "The Development of Our Views Concerning the Nature and Constitution of Radiation" presented in Salzburg at a meeting of German Scientists and Physicians,[3] Planck concluded his relatively long remarks by reproposing his initial interpretation of the radiation energy quanta: ". . . I think that first of all one should attempt to transfer the whole problem of the quantum theory to the area of *interaction* between matter and radiation energy; the processes in pure vacuum could then temporarily be explained with the aid of the Maxwell equations."[4]

The discussion of the fluctuating mirror ends with a remark that Einstein would repeat again and again in the *Notes* and on several different occasions during the last ten years of his life. He realized that his insight into the corpuscular nature of radiation had become the cornerstone of the highly successful theory of quantum mechanics, thus making him a founder of quantum physics, but he remained doubtful about the final status of the contemporary interpretation of that theory: "This dual nature of radiation (and of material corpuscles) is a major property of reality, which has been interpreted by quantum mechanics in an ingenious and amazingly successful fashion. This interpretation, which is looked upon as essentially definitive by almost all contemporary physicists, appears to me to be only a temporary expedient" (*Notes*, p. 49 [p. 171]). Just like Planck, but in an opposite sense, Einstein refers to the present state of understanding as a merely temporary one.

NOTES

1. CPAE vol. 2, Docs. 56, 60.
2. Einstein to von Laue, 17 March 1952, AEA 16-168.
3. CPAE vol. 2, Doc. 60.
4. CPAE vol. 2, Doc. 61, p. 396.

10
THE SPECIAL THEORY
OF RELATIVITY

"THERE IS NO SUCH THING AS SIMULTANEITY OF DISTANT EVENTS"

The type of critical reasoning required for the discovery of this central point was decisively furthered, in my case, especially by the reading of David Hume's and Ernst Mach's philosophical writings.

—Einstein, *Autobiographical Notes*, p. 51 [p. 171]

EINSTEIN WROTE TO HIS FRIEND CONRAD HABICHT AT THE END OF MAY 1905, IN THE letter mentioned previously, that his ideas on the electrodynamics of moving bodies were only roughly drafted. But within about five weeks, they evolved into a groundbreaking paper, "On the Electrodynamics of Moving Bodies," submitted on 30 June to the *Annalen der Physik*.[1] It was the first formulation of the special theory of relativity.

Why did Einstein create special relativity in the first place? Its starting assumption was an extension of the Galilean-Newtonian relativity principle, which stipulates that the laws of mechanics are the same in all inertial frames of reference that move with constant velocity with respect to each other. Einstein extended this principle to all laws of physics. The classical relativity principle can be described by the mental model of a train (with blocked windows) moving at constant velocity. There is no *mechanical* measurement that passengers on that train can perform that would tell them if they are at rest or moving with respect to the platform.

Can this relativity principle be extended to all physical phenomena, including electromagnetic phenomena such as light propagation? According to the then prevailing interpretation of these phenomena, based on Maxwell's equations, this seemed hardly possible. Light was known to be a wave phenomenon, and such phenomena require the existence of a medium to propagate. At the time of emergence of electromagnetism, this medium was called the "ether." The ether was assumed to be immobile and to constitute a preferred system of reference with respect to which the velocity of light is the constant

". . . no such thing as simultaneity of distant events."

explicitly identified in Maxwell's equations. All attempts to detect the existence of the ether by experiments failed. Einstein made the bold assumption, which is incompatible with classical physics, that this velocity is the same in all inertial frames and that one does not have to assume the existence of the ether, thus extending the relativity principle to all physical phenomena. If the velocity of light were not constant, the laws of electromagnetism would be different in different inertial frames.

Several authors have described Einstein's path to the special theory of relativity in the context of his work, which led to the 1905 papers.[2] This is not a simple task, because historians of science studying this process can rely only on scarce contemporary remarks by Einstein and later recollections in correspondence and in autobiographical texts. This is in contrast to general relativity, where we have eight years of extensive correspondence, drafts of calculations, and intermediate publications. In our account of the origin of special relativity, we shall follow mainly the scenario outlined in the *Autobiographical Notes*.

Einstein recalls that shortly after 1900, following Planck's work on black-body radiation, it became clear to him that neither mechanics nor electrodynamics could claim exact validity, except in limiting cases, and that new principles and new laws were required. He describes the emergence of the new laws as a result of an agonizing process: "The longer and the more desperately I tried, the more I came to the conviction that only the discovery of a universal formal principle could lead us to assured results. The example I saw before me was thermodynamics. The general principle was there given in the theorem: The laws of nature are such that it is impossible to construct a *perpetuum mobile* (of the first and second kind). How, then, could such a universal principle be found?" (*Notes*, p. 49 [p. 171]).

This was a very ambitious task for a young man, twenty-six years old, not working at a university, and employed as a full-time clerk at the Swiss patent office in Bern. Thus, his scientific endeavors were mainly confined to his "free time." The year 1905 is known as Einstein's "miraculous year" for his scientific achievement. This term can also apply to the conditions under which these achievements were accomplished. What led him to the discovery of such a principle at a time when the giants of physics like Hendrik Antoon Lorentz and Henri Poincaré, who saw what Einstein saw and contemplated the same issues, did not even realize that a new principle was necessary? They still tried to cope with the problems in the framework of classical mechanics of Galileo and Newton and classical electrodynamics of Maxwell and Lorentz. Einstein gives a partial answer to this question, tracing the origin of his success to the time when he was sixteen years old: "After ten years of reflection such a principle resulted from a paradox upon which I had already hit at the age of sixteen" (Notes, p. 49 [p. 171]). At that age, young Albert asked himself how a light wave would look to an observer moving alongside this wave at the speed of light. He would have to see the oscillating electric and magnetic fields, constituting an electromagnetic wave, but stationary in space. It seemed, however, that there was no such a thing. This thought experiment also raised the question of what speed of light would an observer, moving along a light wave with a given velocity, measure. The answer depended on the underlying model of the ether, the hypothetical medium carrying light waves. In a stationary ether, which is not carried along by the moving system, the speed of light relative to the observer would certainly have to change depending on his state of motion. It is exactly such changes that the Michelson-Morley experiment was designed to detect but did not find (see box). It is not clear if Einstein knew about this experiment at the time of writing his paper in 1905. He does not clarify this uncertainty in the Notes. There he claims that it was clear to him intuitively, already at an early stage, that everything, including the speed of light, would look to the moving observer the same as it does to an observer at rest relative to the earth.

THE MICHELSON-MORLEY EXPERIMENT

The Michelson–Morley experiment was performed by Albert A. Michelson and Edward W. Morley in 1887 in Cleveland, Ohio. It was designed to measure the difference between the speed of light propagating in the direction of the earth's rotation and in the perpendicular direction. The result was negative, in that no such expected difference was found to exist. This result is generally considered to be the first experimental evidence against the existence of ether as the "light-medium" in which light propagates and as the preferred reference frame with respect to which its speed is measured. Einstein did not mention the Michelson-Morley experiment explicitly in his first paper on special relativity, but he seems to allude to it saying that ". . . the unsuccessful attempt to detect a motion of the earth relative to the 'light-medium'" was among the causes that led him to the conclusion ". . . that not only the phenomena of mechanics but also those of electrodynamics have no properties that correspond to the concept of absolute rest." From a letter to Mileva Marić, we know that he had read by that time a review article by Wilhelm Wien on numerous experiments testing the motion of the earth with respect to the ether.[3]

Although not mentioned in the Notes, we know that there was another puzzle that challenged the young Einstein. He found it curious that the basic phenomenon of electrodynamics describing the interaction of electric charges in a conductor with a magnet is formulated by two different laws. One law refers to the case of a moving conductor and a magnet at rest; in the other case, the magnet is moving with respect to a conductor at rest. The result is the same in the two cases. The independence of the interaction between magnet and conductor from the state of the observer would be an immediate consequence of the principle of relativity in mechanics, if only this principle were also valid for electrodynamics. But this seemed to be excluded as the hypothetical ether, on which contemporary electrodynamics was based, constituted a preferred system of reference. This elementary argument is mentioned in the introductory remarks in the paper "On the Electrodynamics of Moving Bodies" as one of the incentives for the special theory of relativity.

Einstein does not tell us, in the *Notes*, what attempts he made to resolve the puzzles that troubled him for so long.[4] In 1905, he eventually realized that the concept of time is not simply given but represents a rather complicated construct depending on a measuring method involving the synchronization of clocks. This conclusion enabled him to question the absolute character of time, as it underlies classical physics and it resonated with his reading of the philosophical writings of Ernst Mach and David Hume.

Einstein realized that a new understanding of the meaning of space and time measurements was necessary. In particular, methods of measuring a distance in space and a duration in time between two events had to be defined, taking into account the need to relate such measurements performed in frames of reference in relative motion to each other. To this end, he introduced the concept of measuring rods and clocks and addressed such questions as: How do measuring rods and clocks behave in inertial frames in relative motion to each other? What does it mean to say "this event takes place simultaneously with another one," and how does one check that?

The discussion of the meaning of time in Einstein's paper "On the Electrodynamics of Moving Bodies" begins with a detailed analysis of the meaning of the statement "the train arrives here at 7 o'clock." A reasonable interpretation of this statement is to say that the small hand of the watch of an observer standing on the platform points at the number 7 simultaneously with the arrival of the train. This defines the time of an event, as the time measured by a clock, located at the place of the event, but this definition is not satisfactory when the time of events occurring at different positions has to be compared. The analysis of such a comparison requires a careful examination of the meaning of time and its measurement, and the principle of the constancy of the speed of light leads to the conclusion that the basic notion of simultaneity must be relative to the reference frame chosen. If an observer in one inertial frame of reference decides, by a well-defined method, that two events occur simultaneously, then these two events will not be simultaneous in another frame of reference moving with constant velocity with respect to the first one. A new interpretation of the concept of simultaneity is required because of the finiteness of the speed of light through which observers receive information about events at distant locations.

The relative character of time is essential to resolve the paradox implied by the thought experiment of the sixteen-year-old Einstein. This paradox is formulated, in the *Notes*, as the contradiction between the two basic assumptions: the principle of the constancy of the speed of light and the principle of relativity, stating that the laws of physics are the same in all inertial frames. Each one of these principles is supported by experience, but, in Newtonian physics, they are mutually incompatible, because the relation between the spatial and temporal coordinates describing a specific event in two different inertial frames violates the principle of the constancy of the speed of light.

The fundamental insight that led Einstein to the special theory of relativity was the realization that these two assumptions are compatible if the coordinates describing a specific event in different inertial frames are related by the so-called Lorentz transformation. An event is defined by the three coordinates x, y, z, defining its position in space, and by the time coordinate t. In another inertial frame, the space and time coordinates are x´, y´, z´, and t´. In Newtonian physics t = t´. In the special theory of relativity, t´ is a function of x, y, z, t. The Lorentz transformation is a mathematical expression of the dependence of x´, y´, z´, t´ on x, y, z, t. It was discovered by Lorentz as the transformation that guarantees Maxwell's equations are the same in all inertial frames. The electric and magnetic fields in these equations are functions of the space and time coordinates. In different inertial frames they assume different values, but they are related by the Lorentz transformation, which assures that the equations themselves are invariant.

In Lorentz's theory, this transformation made sure that the electrodynamics of moving bodies complied with all measurements showing that motion with regard to the ether has no observable effects. They thus played an auxiliary role. In contrast, in Einstein's formulation of the special theory of relativity, the Lorentz transformation assumes a profoundly different meaning. It is a property of the four-dimensional space-time continuum. It defines the acceptable laws of nature, which are the ones that are invariant under this transformation. This is the universal principle for which Einstein was looking for ten years, comparable to the principle of nonexistence of the *perpetuum mobile* in thermodynamics. Both are restrictive principles for admissible laws and processes in nature.

Einstein devotes a paragraph to a discussion of the concept of four-dimensional space-time introduced by the mathematician Hermann Minkowski. He was a professor of mathematics at the Federal Technical University in Zurich when Einstein was a student there, and Einstein attended several of his courses. In 1908, he showed that Einstein's special theory of relativity could be understood geometrically as a theory of four-dimensional space-time. In classical physics time is absolute and, therefore, there is no advantage in treating "events" described by the four parameters x,y,z,t as points in a four-dimensional space-time. Instead, we have a three-dimensional space continuum and an independent one-dimensional time continuum. The situation is different in special relativity. Since the time *t´* of an event observed from another inertial system depends on both its time and space coordinates in the initial system, it is this mixing of spatial and time coordinates that makes it convenient to combine them into a single four-dimensional space-time.

Minkowski's four-dimensional space-time is equipped with a "metric" instruction that is employed to measure the distance between two events. The square of this distance is simply the square of the time separation between the two events (multiplied by the velocity of light squared) minus the square of their spatial separation. This is essentially the Pythagoras theorem in space-time. This "distance" between two events is invariant under Lorentz transformations between different inertial coordinate systems.

It took Einstein some time to appreciate Minkowski's geometric formulation of the theory of special relativity as an interesting and useful contribution. He became convinced of its basic importance only around 1912, during his search for a relativistic theory of gravitation. Minkowski's formulation of the theory became the framework for its later development and led Einstein to his theory of general relativity. In the first paragraph of Einstein's seminal article "The Foundation of General Relativity," published in March 1916, he writes: "The generalization of the theory of relativity has been facilitated considerably by Minkowski, a mathematician who was the first one to recognize the formal equivalence of space coordinates and the time coordinate, and utilized this in the construction of the theory."[5]

Einstein concludes the discussion of the special theory of relativity with the insights that this theory provided for physics. He argues that the statement that "there is no such thing as simultaneity of distant events" implies that action between distant points has to be mediated by continuous functions in space (fields) and that "the material point, therefore, can hardly be retained as a basic concept of the theory" (*Notes*, p. 57 [p. 173]). This is a profound, far-reaching consequence, which will be discussed again in the context of Einstein's quest for the unified field theory. In the *Notes*, Einstein mentions only briefly, in passing, what is probably the best-known result of the theory, the equivalence of mass and energy described by the famous equation $E = mc^2$, which overcomes the notion of mass as an independent concept (*Notes*, p. 57 [p. 173]).

The *Autobiographical Notes* makes it clear that the genesis of special relativity was of a special character. It did not emerge, as the other topics of Einstein's "miraculous year," from taking up a challenging problem of contemporary physics. The genesis of special relativity was different, reaching back to the earliest scientific ideas of the adolescent Einstein, accompanying his pathway into physics as a leitmotif. The wrestling with the notions of space and time was, for Einstein, a drama that unfolded on a time scale of its own, beginning when he was sixteen and by far not terminated even after he completed general relativity at the age of thirty-six.

NOTES

1. Reprinted and discussed in John Stachel, ed., *Einstein's Miraculous Year: Five Papers That Changed the Face of Physics* (Princeton, NJ: Princeton University Press, 2005). Originally published 1998.
2. See, for example, Jürgen Renn and Robert Rynasiewicz, "Einstein's Copernican Revolution," and John Norton, "Einstein's Special Theory of Relativity and the Problems of Electrodynamics That Led Him to It," both in *The Cambridge Companion to Einstein*, ed. Michel Janssen and Christoph Lehner (Cambridge: Cambridge University Press, 2014).
3. Einstein to Marić, 28? September 1899, in Jürgen Renn and Robert Schulmann, eds., *Albert Einstein—Mileva Marić: The Love Letters* (Princeton, NJ: Princeton University Press, 1992) 15.
4. On this point, see, for example, Renn and Rynasiewicz, "Einstein's Copernican Revolution."
5. Albert Einstein, "The Foundation of General Relativity," *Annalen der Physik* 49 (1916): 769–822; reprinted in CPAE vol. 7, Doc. 30.

11

THE GENERAL THEORY
OF RELATIVITY

"WHY WERE ANOTHER SEVEN YEARS REQUIRED?"

The acceleration of a system falling freely in a given gravitational field
is independent of the nature of the falling system. . . . It turned out
that, within the framework of the program sketched, this simple state
of affairs could not at all, or at any rate not in any natural fashion,
be represented in a satisfactory way. This convinced me that, within
the structure of the special theory of relativity there is no niche for a
satisfactory theory of gravitation.

—Einstein, *Autobiographical Notes*, p. 61 [pp. 174–175]

THE FIRST CHALLENGE OF THE SPECIAL THEORY OF RELATIVITY WAS TO INCORPORATE
gravitation, that is, the force of gravity between two masses, into the framework of the
new theory. This turned out to be a difficult task, because Newton's law of gravity assumes
an instantaneous action at a distance. This law, in its classical form, was not compatible with the special theory of relativity, which requires that any interaction between two
objects cannot propagate faster than the velocity of light. This, however, was not the only
difficulty.

The gravitational field has a peculiar property. Unlike the case of an electric or magnetic field, in a gravitational field, bodies of any size or material composition, starting
from rest or uniform motion, will move at the same acceleration. This is one of the basic
principles of classical physics, established by Galileo in his lifelong studies of falling bodies and, mythically, confirmed by him by dropping objects from the tilted tower in Pisa.
This principle implies that the *inertial mass* of a body is always equal to its *gravitational
mass*, although conceptually the two masses are distinct. The inertial mass determines the
acceleration of a body caused by a given force, while the gravitational mass determines
the force exerted on a body by a given gravitational field. The equivalence of these two
properties of a massive body was known in mechanics, and its validity had already been

"Why were another
seven years required?"

demonstrated empirically with great accuracy in Einstein's time, but its significance had
not been explored. Only Einstein interpreted it as a basic principle and adopted it as a
cornerstone of his general theory of relativity.

From the equality of the inertial and gravitational mass of a body Einstein deduced his
famous "equivalence principle." An observer in a closed box in outer space, far away from
all heavenly bodies, will not feel a gravitational field in his or her surroundings. Suppose
that the box is made to move "upward" with constant acceleration. The person in the box
has no way to decide if the effects he or she observes in the box are caused by a uniform
acceleration of the box or by a gravitational field exerting a gravitational force in the
opposite direction. Likewise, the passenger on a train who feels a tilt backward when the
train is suddenly accelerated may assume that the train is at rest but that a gravitational
field has suddenly been applied to the system. Later in his life, Einstein referred to this
revelation as "the happiest thought" of his life.[1]

There was an obvious way to make classical gravitational theory compatible with the
principles of the special theory of relativity, and Einstein was initially thinking in this
direction. However, the problem with this obvious generalization was that the resulting
theory of gravitation seemed to violate Galileo's principle that all bodies fall with equal
acceleration. The relation between inertial mass and energy in special relativity, expressed
by the formula $E = mc^2$, must imply that in a relativistic theory of gravitation, the gravitational mass of a physical system should also depend on the energy in a precisely known
way so as to maintain Galileo's principle. Contemporary scientists such as Max Abraham
and Gustav Mie, for instance, were quite ready to abandon Galileo's principle in order to
obtain a relativistic theory of gravitation in the sense of special relativity. For Einstein,
this was a basic principle of physics, and if a theory did not achieve this in a natural way,
it was to be abandoned.

The obvious program of developing a new theory of gravitation within the framework of the special theory of relativity led him to the conclusion summarized in the epigraph to this chapter. Thus, a new theory of gravitation was needed, but it was not clear how such a theory should look, what heuristic assumptions could be made, and even what specific criteria it should satisfy.

The failure to fit Newton's well-established law of gravity into the framework of the special theory of relativity led Einstein to question its concepts of space and time and caused him to continue the revolution with his 1915 theory of general relativity. A natural formulation of a *general* principle of relativity would read: all frames of reference are equivalent for the description of the laws of nature, whatever may be their state of motion. Such a generalization seemed to Einstein to be an intellectual necessity. In his popular book *Relativity: The Special and the General Theory*, he writes: "since the introduction of the special principle of relativity has been justified, every intellect which strives after generalization must feel the temptation to venture the step towards the general principle of relativity."[2]

The equivalence principle states that all physical processes in a uniform and homogeneous gravitational field are equivalent to those that occur in a uniformly accelerated frame of reference without a gravitational field. This implies that the laws of physics have to be the same in the two systems. Thus, the "Lorentz transformation," which assures the invariance of physical laws between all inertial frames is too narrow. Einstein concluded that their invariance must be postulated also with respect to more general transformations. The spatial and temporal coordinates x′, y′, z′, t′ of an event in one reference frame can now be nonlinear functions of the coordinates x, y, z, t in another one.

Einstein recalls that this was known to him already in 1908 and poses the question why another seven years were required for the completion of the general theory of relativity in November 1915. Actually, the conclusions of the preceding paragraph were already exposed in 1907, in a review article on the special theory of relativity, when Einstein first conceived the idea of a relativistic theory of gravitation on the basis of the principle of equivalence. In that article he also showed that this principle implied the bending of light by gravitation and the dependence of the rate of clocks on the gravitational field at their location. Four years later, when he was serving as professor of theoretical physics in Prague, he gave a more complete formulation of these implications of the principle of equivalence.

Let us mention, in passing, that the *Autobiographical Notes* makes no geographical references, in the sense that places like Prague, Zurich, Berlin, or Princeton are not mentioned. Banesh Hoffmann, who collaborated with Einstein in Princeton, and Helen Dukas, Einstein's devoted secretary, point out in their discussion of the *Autobiographical Notes* that: "The 'Notes,' though autobiographical, are in no sense geographical. They are essentially placeless. Wherever he went, his ideas went with him, and where he went was here irrelevant. The 'Notes,' are not wholly placeless, though. They tell of a unique adventure—and a world-shaking one—that took place within the ivory tower of a mind."[3]

The *Autobiographical Notes* is indeed not geographical, yet, the transition from the special to the general theory was a process closely associated with particular places. In our book *The Road to Relativity*,[4] we describe the eight years that followed the first formulation of the equivalence principle in Bern as a tale of three cities: Prague, Zurich, and Berlin. Each one provided a different social and political environment, and each one was related to a different phase in his family life. Gerald Holton, a pioneer of Einstein scholarship in the historical and philosophical context, addressed the question of whether Einstein could have developed his general theory of relativity anywhere else than in Berlin and went as far as claiming: "No other man than Einstein could have produced General Relativity, and in no other city than in Berlin."[5]

Einstein soon realized that the equivalence of all reference frames implies a non-Euclidean or curved space-time. We can visually imagine a curved surface but not a curved four-dimensional space-time. In fact, Einstein had encountered non-Euclidean geometry as a student in the form of Gauss's theory of curved surfaces formulated in 1828. About twenty-five years later, this theory had been extended to arbitrary dimensions by the mathematician Bernhard Riemann, but these more sophisticated methods, which turned out to be of crucial importance to the new theory of gravitation, were unfamiliar to most physicists, including Einstein. Therefore, in 1912, Einstein solicited the help of his friend from student years, the mathematician Marcel Grossmann, to help him with the mathematics. In ordinary, Euclidean space, the distance between two points is determined by their coordinates. In special relativity, the distance between two events is determined by the spatial and temporal coordinates of the two events. This is no longer the case in a curved space-time.

The distance between two points depends on the structure of space-time in their neighborhood and not only on their coordinates. In curved four-dimensional space-time, ten numbers are needed to calculate the distance from one point to any neighboring point. It is convenient to present these numbers in a 4×4 matrix array g_{ik}, where the first index ($i = 1,2,3,4$) represents the line, and the second ($k = 1,2,3,4$), the column of this matrix. This array is the "metric tensor," which reflects the geometric properties of space-time in a chosen coordinate system (see *Notes*, eq. 2, p. 67 [p. 176]). Its components, in general, are functions of position in space-time. The metric tensor has sixteen components, but only ten of them are independent, because of the symmetry between the off-diagonal elements, $g_{12} = g_{21}$, ... Einstein soon realized that these ten functions are at the same time components of the gravitational potential. Thus, in general relativity, the structure of space-time and the gravitational field are represented by the same mathematical entity.

Because in general relativity the coordinates do not determine the distance between two events, they lose their physical meaning. In the *Notes*, Einstein mentions the difficulty in adapting to this notion as the main reason for the long duration of the process to general relativity. This, however, is not an accurate account of what really happened. The equivalence of the different coordinate systems implies that the laws of physics are invariant under all continuous transformations of coordinates. This is the "principle of general covariance," and Einstein was looking for general covariant field equations. Actually, by the end of 1912 he already had the essentially correct formulation of the theory that he presented in November 1915. He misinterpreted the result and abandoned it. Instead, he and Grossmann published a theory that did not meet this basic goal. It was not "generally covariant." Over three years, Einstein proposed one argument after another to convince himself and the science community that this was the best that could be done.

There is no trace in the *Notes* of this convoluted process. It seems that the main goal of Einstein's discussion of the transition from the special to the general theory of relativity is to prepare the ground for the search for the unified field theory (see chapter 13) and to indicate that he had always treated his general theory of relativity as incomplete.

Einstein assumed that the generalization of the nature of the physical space in the field-free case of special relativity involved two steps:

(a) the pure gravitational field
(b) the general field (which is also to include quantities that somehow correspond to the electromagnetic field) (*Notes*, p. 69 [p. 177])

This statement is followed by a thought-provoking remark: "It seemed hopeless to me at that time to venture the attempt of representing the total field (b) and to ascertain field laws for it. I preferred, therefore, to set up a preliminary formal frame for the representation of the entire physical reality; . . ." (*Notes*, p. 69 [p. 177]).

Did he really have the second step in mind as early as 1915 and consider his achievement as setting the stage for the inclusion of the electromagnetic field at a later stage? In the correspondence with colleagues, following the final formulation of the general theory of relativity, in which he shared with them his joy and satisfaction, there was no sign of that. However, a few years later it is clearly expressed, for example, in his Nobel Prize lecture (see chapter 13).

The triumphal achievement of Einstein's effort in 1915 was the gravitational field equation (presented in the *Notes* on p. 71). The left-hand side of this equation represents the gravitational field given by the geometrical structure of space-time and the right-hand side represents the energy and momentum density acting as the source of the field. Einstein claims that he was aware of the provisional character of the right-hand side: "Not for a moment, of course, did I doubt that this formulation was merely a makeshift in order to give the general principle of relativity a preliminary closed form expression. For it was essentially *no more* than a theory of the gravitational field, which was isolated somewhat artificially from a total field of as yet unknown structure" (*Notes*, p. 71 [p. 178]).

Years later, in 1936, Einstein described this equation as follows: "The theory . . . is similar to a building, one wing of which is made of fine marble (left part of the equation), but the other wing of which is built of low-grade wood (right side of the equation). The phenomenological representation of matter is, in fact, only a crude substitute for a representation which would do justice to all known properties of matter."[6]

There are no such explicit remarks in Einstein's seminal paper "The Foundation of General Relativity," where this equation was first derived. Einstein's account of his road to general relativity is clearly influenced by his interests at the time of writing the *Notes*. Moreover, the remarks on the nature of the theory and on its basic features are not reminiscences. They are part of his daily struggles in search for a unified field theory.

NOTES

1. "der glücklichste Gedanke meines Lebens," CPAE vol. 7, Doc. 31, p. 136.
2. Albert Einstein, *Relativity: The Special and the General Theory; 100th Anniversary Edition*, ed. Hanoch Gutfreund and Jürgen Renn (Princeton, NJ: Princeton University Press, 2015), 74.
3. Banesh Hoffmann and Helen Dukas, *Albert Einstein: Creator and Rebel* (New York: Plume Books, 1972), 12.

4. Hanoch Gutfreund and Jürgen Renn, *The Road to Relativity: The History and Meaning of Einstein's "The Foundation of General Relativity," Featuring the Original Manuscript of Einstein's Masterpiece* (Princeton, NJ: Princeton University Press, 2015).

5. Gerald Holton, "Who Was Einstein? Why Is He Still So Alive?," in *Einstein for the 21st Century: His Legacy in Science, Art, and Modern Culture*, ed. Peter L. Galison, Gerald Holton, and Silvan S. Schweber (Princeton, NJ: Princeton University Press, 2008), 4.

6. Einstein, "Physics and Reality" (1936), reprinted in *Ideas and Opinions: Based on "Mein Weltbild,"* ed. Carl Seelig (New York: Bonanza Books, 1954), 311.

1 2
QUANTUM MECHANICS

"THIS THEORY OFFERS NO USEFUL POINT OF DEPARTURE FOR FUTURE DEVELOPMENT"

> Before I enter upon the question of the completion of the general theory
> of relativity, I must take a stand with reference to the most successful
> physical theory of our period, viz., the statistical quantum theory which
> assumed a consistent logical form about twenty-five years ago.
>
> —Einstein, *Autobiographical Notes*, p. 77 [p. 179]

HAVING COMPLETED THE DISCUSSION OF HIS ROAD TO GENERAL RELATIVITY AND ITS basic features, Einstein begins to lay the groundwork for an extension of this theory toward a theory of "the total field," namely, the unified theory (*Notes*, p. 69 [p. 177]). He interrupts this exposition, declaring that he now wishes to express his views on "the most successful physical theory of our period, viz., the statistical quantum theory" (*Notes*, p. 77 [p. 179]). He consistently refers to quantum theory as a statistical theory to emphasize its probabilistic nature. The presentation from this point on is entirely in the present tense— not what he thought about it in the past, as would be appropriate for a comprehensive autobiographical text, but what he thinks about it now. Thus, there is no reference to his debates with Niels Bohr in the 1920s. The major challenge standing before the theoretical physics community, already at the time of Einstein's writing the *Notes*, was how to combine the quantum theory and the theory of relativity into a single theory of the physical reality. All the attempts to achieve this goal had not succeeded, and the question remained about the theoretical foundation of physics in the future: "Will it be a field theory? Will it be in essence a statistical theory?" (*Notes*, p. 77 [p. 179]).

For most physicists, the quantum theory was a final theory and, therefore, a cornerstone of every future comprehensive theory of the physical world. Einstein's opinion was diametrically opposite: "It is my opinion that the contemporary quantum theory represents an optimum formulation of the relationships, given certain fixed basic concepts, which by and large have been taken from classical mechanics. I believe, however, that this theory offers no useful point of departure for future development" (*Notes*, p. 83 [p. 181]).

"... no useful point of departure for future development."

He believed that a future theory, based on the extension of the field theory of general relativity, that is, the second step of his generalization of the special theory of relativity, would provide a substitute for the statistical theory of quantum mechanics as the ultimate theory of matter. This is why he interrupts his explanation with a digression on quantum mechanics to explain why he considers it, unlike general relativity, not as an intermediate step on the road to a final theory, but as a dead end. We shall return to this point in the next chapter.

In his essay "Physics and Reality" (1936), Einstein highlights the success of quantum mechanics and thus indicates the magnitude of the challenge to replace it with another theory: "Probably never before has a theory been evolved which has given a key to the interpretation and calculation of such a heterogeneous group of phenomena of experience as has quantum theory. In spite of this, however, I believe that the theory is apt to beguile us into error in our search for a uniform basis for physics, because in my belief, it is an *incomplete* representation of real things. . . . The incompleteness of the representation leads necessarily to the statistical nature (incompleteness) of the laws."[1] In the *Notes*, Einstein engages in a detailed explanation of his reasons for this opinion.[2]

In Schrödinger's formulation of quantum mechanics, the state of a system (for example, a particle) is characterized by a "wave function" ψ, which is a function of parameters like the position (q) and momentum (p) of a particle and time. According to a rule formulated by the physicist Max Born, the function ψ can be interpreted as predicting the probability of finding a definite value of the parameter q (or p) at a given time. This probability can also be determined empirically by preparing the same state many times and averaging over the results of many measurements of that parameter. Let us look at two ways to interpret the result of a single measurement of, say, q. One possibility is that the value measured is the value of that parameter before the measurement. In that case, the function ψ is not a complete description of the system because it only tells us what we know from many measurements. The other possibility is that the measured value, implied by the ψ-function, is produced by the measurement itself. In that case, the function ψ describes the system completely.

The difference between these two possibilities was at the core of the grand debate between Einstein and Niels Bohr in the 1920s, particularly at the Solvay conference in 1927.[3] The second possibility is the basis of the Copenhagen interpretation of quantum mechanics.[4] Einstein did not accept the idea that there is no objective reality, independent of an observation. He would not abandon the causal nature of classical mechanics and of the field theories of electromagnetism and gravitation and accept the probabilistic character of quantum theory as the last word. He presented one argument after another, invented one thought experiment after another to challenge the validity or the consistency of the theory. Bohr disputed every argument, but Einstein remained unconvinced. At a later stage, instead of claiming that the theory is wrong, he argued that it is incomplete and that it will be replaced by a comprehensive causal theory in the future. The essence of this argument is represented in the—by now famous—Einstein-Podolsky-Rosen paper, published in 1935.[5] This paper has been depicted as the "EPR paradox." The authors argued that the second meaning of the ψ-function (as defined in the preceding paragraph) leads to a paradox, and therefore the ψ-function cannot provide a complete description of the physical reality.

The line of thought presented in the *Notes* is based on this "paradox," without mentioning its name explicitly. It describes a system that has been separated into two subsystems, so far apart that they cannot influence each other. Such an influence would imply that information is transmitted faster than light, violating a basic principle of relativity. The entire system is described by a function ψ, and a measurement performed on one subsystem determines the result of the second subsystem, without any uncertainty. Einstein concludes that the values of the parameters of the second subsystem are, in reality, precisely defined before the measurement, because there is no way that the measurement of the first subsystem can affect the other system, unless one accepts a "spooky action at a distance." Einstein's conclusion is that the ψ-function cannot be a complete description of the physical reality, because depending on the result of measurements performed on one system, different ψ-functions would describe the same factual reality of the other system.

Einstein concludes the presentation of his "stand with reference to the most successful theory" of his period, by stating: "The statistical character of the present theory would then follow necessarily from the incompleteness of the description of the systems in quantum mechanics, and there would no longer exist any ground for the assumption that a future foundation of physics must be based upon statistics" (*Notes*, p. 81 [p. 181]). Einstein expressed his conviction that the future of physics would eventually take this course in his "Reply to Criticisms": "Assuming the success of efforts to accomplish a complete physical description, the statistical quantum theory would, within the framework of future physics, take an approximately analogous position to the statistical mechanics within the framework of classical mechanics. I am rather firmly convinced that the development of theoretical physics will be of this type; but the path will be lengthy and difficult."[6] Although not mentioned explicitly, this statement has been quoted as clear evidence of Einstein's commitment to the notion of "hidden variables."[7] The discussion of the Einstein-Podolsky-Rosen paradox led to the formulation and deeper understanding of a basic feature of quantum mechanics—quantum entanglement. The further exploration of the apparent paradox eventually even helped to bring about the new paradigm of quantum information science, which in turn lays the groundwork for such promising technologies as quantum computing, quantum communication, and quantum cryptography.

NOTES

1. Einstein, "Physics and Reality" (1936), reprinted in *Ideas and Opinions: Based on "Mein Weltbild,"* ed. Carl Seelig (New York: Bonanza Books, 1954), 315–316.
2. For Einstein's views on quantum mechanics, see also Christoph Lehner, "Realism and Einstein's Critique of Quantum Mechanics," in *The Cambridge Companion to Einstein*, ed. Michel Janssen and Christoph Lehner (Cambridge: Cambridge University Press, 2014).
3. Guido Bacciagaluppi and Antony Valentini, *Quantum Theory at the Crossroads: Reconsidering the 1927 Solvay Conference* (Cambridge: Cambridge University Press, 2009).
4. See Mara Beller, *Quantum Dialogue: The Making of a Revolution* (Chicago: University of Chicago Press, 1999).
5. A. Einstein, B. Podolsky, and N. Rosen, "Can Quantum-Mechanical Description of Physical Reality Be Considered Complete?" *Physical Review* 47, no. 10 (1935): 777–780.
6. Schilpp, *Albert Einstein: Philosopher-Scientist*, 672.
7 J. S. Bell, *Speakable and Unspeakable in Quantum Mechanics* (Cambridge: Cambridge University Press, 1987), 89.

13
THE UNIFIED FIELD THEORY

"FINDING THE FIELD EQUATIONS
FOR THE TOTAL FIELD"

. . . it would be most beautiful if one were to succeed in expanding the
group once more in analogy to the step that led from special relativity
to general relativity.

—Einstein, *Autobiographical Notes*, p. 85 [p. 182]

THE CHALLENGE OF COMBINING GRAVITATIONAL AND ELECTROMAGNETIC FIELDS IN A
single theoretical framework was addressed even before the completion of general relativity by the mathematician David Hilbert. He attempted to integrate Einstcin's theory
of gravitation with the nonlinear theory of electromagnetism proposed by Gustav Mie

Bringing all ends together.

(see box below). At the time, Einstein considered this attempt as naive. On the other hand, Einstein himself had already alluded in his first comprehensive summary of the general theory of relativity, published in 1916, to the challenge of bringing electromagnetism and gravitation into the framework of a single theory. He left open the question whether the combination of gravitation and electromagnetism would lead to a new theory of matter, which Gustav Mie had tried to build on electromagnetism alone: "In particular it may remain an open question whether the theory of the electromagnetic field in conjunction with that of a gravitational field furnishes a sufficient basis for the theory of matter or not. The general postulate of relativity is unable in principle to tell us anything about this. It must remain to be seen, during the working out of the theory, whether electromagnetics and the doctrine of gravitation are able in collaboration to perform what the former by itself is unable to do."[1]

GUSTAV MIE'S THEORY OF MATTER

The German physicist Gustav Mie (1869–1957) proposed, in 1912, a theory of matter based on a nonlinear extension of Maxwell's electrodynamics in the framework of the special theory of relativity. His hope was to explain the elementary particles known at that time, electrons and protons, as emerging properties of a universal electromagnetic field, thereby overcoming the conceptual duality between field and matter. His idea was to look for a nonlinear formulation of Maxwell's equations that would have solutions with a very high intensity around a certain point in space. With an appropriate equation of motion, such a region in space can then be interpreted as a particle. He also tried to include gravitation in his electrodynamic theory, with the quest for a unified theory of physics.

Shortly after the publication of Einstein's theory of general relativity, the prospect of unifying gravitation and electromagnetism generated wider interest in the physics community. The pioneers of this effort were Hermann Weyl, Theodor Kaluza, and Arthur Eddington. Initially, Einstein's involvement in this endeavor was confined to reacting to the work of others through correspondence and comments published mainly in the *Proceedings of the Prussian Academy of Sciences*. His own comprehensive proposals on this subject started only in 1925 and persisted until the end of his life, when most of the other players shifted to other topics, specifically, to quantum mechanics.

Einstein's theory of gravitation was perceived by himself, and by others, as partial and incomplete, to be supplemented by a unified theory, which would include the electromagnetic field on the same footing. In his Nobel Prize address, delivered to the Nordic Assembly of Naturalists at Gothenburg in July 1923 (Einstein was not present at the Nobel Prize award ceremony in 1922), Einstein defined this as a goal to be achieved: "A mind seeking unity within the theory cannot be content with the existence of two, in essence entirely independent fields. A mathematically unified field theory is sought in which the gravitational field and the electromagnetic field are conceived as merely different components or manifestations of the same unified field, and where the field equations may no longer consist of logically mutually independent summands."[2]

We have already mentioned the dualistic nature of the Maxwell-Lorentz theory of electromagnetism represented by the coexistence of fields and material particles. In this sense the theory of gravitation was also a dualistic theory. In both theories, the sources of the field were charges and massive particles. The motion of these particles and charges was described—or so it seemed—in both theories by equations of motion independent and in addition to the field equations. The latter point underwent, however, a serious revision when it was realized that there is a basic difference between theories based on linear and nonlinear field equations. Maxwell's field equations are relations between the spatial and temporal rates of change of the fields (partial derivatives of the fields) and not their second or higher powers. In such a linear field theory, the motion of material particles is determined by equations of motion that are not implied by the field equations. On the other hand, in Einstein's nonlinear theory of general relativity, the equations of motion of particles in a gravitational field can be deduced from the field equations themselves. Such equations of motions were first derived in 1927 by Einstein and Jakob Grommer, and in 1938 by Einstein, Leopold Infeld, and Banesh Hoffmann (known as the EIH equations).[3] Without mentioning these works explicitly, the importance of this conclusion is emphasized in the *Notes*: ". . . it turned out that the law of motion need not (and must not) be assumed independently, but that it is already implicitly contained within the law of the gravitational field" (p. 75 [p. 179]). This insight may have been a turning point in Einstein's thinking in the course of his unification program. We shall come back to this point later.

Thus, the goal of unification posed a double challenge—to combine gravitation and electromagnetism into one field, represented by the geometry of space-time, in which, furthermore, matter and its motion would be deduced from the field equations themselves. Such a unification scheme was expected to also account for the basic properties of matter, in particular, for the nature of the two known particles in those days—the electron and the proton. A theory of this kind would explain the microscopic properties of elementary particles, the constituents of atoms, and at the same time would account for the macroscopic phenomena constituting the fabric of the universe. These were the goals pursued by Einstein and, particularly, in the earlier years, by a number of his contemporaries.

In 1925, Einstein published his first original attempt to unify the gravitational and electromagnetic fields. The metric tensor was assumed to be asymmetric. Einstein failed, however, in his attempt to carry this program to a physically sound theory. He abandoned it and for twenty years explored other ways to achieve the goal of unification. He believed it was the strategy of pursuing mathematical formulations with a quest for simplicity that had led him finally to his successful formulation of general relativity. Based on this view, he continually tried new mathematical approaches, and when they did not produce the expected results, he discarded them.[4]

Around 1945, Einstein returned to the approach he had tried twenty years earlier. Again, it was based on an asymmetric metric tensor in which all the sixteen elements are independent functions of space-time coordinates. Ten of them, or rather ten combinations of them, would represent the gravitational field as in the theory of general relativity, and the other six were expected to represent electromagnetism, which is characterized by the six components of the electric and magnetic vector fields. Einstein pursued this approach during the last ten years of his life.

To his longtime colleague Erwin Schrödinger, Einstein sent his first attempts, along these lines: "I am sending you herewith the two papers. . . . I am sending them to no one else because you are the only person I know who is not wearing blinders in regard to the fundamental questions in our science. The attempt depends on an idea that at first seems antiquated and trivial: the introduction of a nonsymmetric tensor g_{ik} as the only physically relevant field quantities."[5]

Schrödinger responded with two detailed letters posing questions and raising critical remarks on Einstein's theory. Einstein was very grateful and encouraged by Schrödinger's interest in his work: "I am very thrilled that you responded in such detail to my new hobby and in such a remarkably short time. It is really admirable. If I were a truly respectable person, I would simply thank you and leave you in peace. But I am not capable of doing so and instead *must* somehow answer your remarks."[6] These remarks and responses evolved during the following months into an intensive correspondence with more than a dozen letters exchanged between the two colleagues.

Struggling with the difficulties of his undertaking and frustrated with the lack of tangible progress, Einstein wrote to Schrödinger: "How well I understand your hesitation! I must tell you right away that deep inside I am not so certain as I previously asserted. . . . We have wasted a lot of time on this and the result looks like a gift from the devil's grandmother."[7] We quote this correspondence because it took place exactly at the time when Einstein wrote his *Autobiographical Notes*.

Let us mention two additional points that guided Einstein in his thinking at that time, which are mentioned in the *Notes*. They are concerned with the invariance of the physical laws under transformations from one reference frame to another and with the representation of material particles in the solutions of the field equations.

The Lorentz transformation between two inertial coordinate frames essentially determines the structure of Maxwell's equations. All the transformations form a mathematical entity called a "group" (the main, but not only, feature that makes it a group is that one Lorentz transformation, followed by another one, forms also a Lorentz transformation). The transition to general relativity involved a wider group of transformations. Einstein believed that this, together with the symmetric metric tensor, almost determines the gravitational field equations. Guided by this insight, he naturally concluded that it would be beautiful if one could widen the group of transformations once more. All his attempts in this direction were, however, unsuccessful, and he concluded that the most satisfactory approach was to limit oneself to the continuous transformation of coordinates, as in general relativity, and to assume a nonsymmetric metric tensor.

The other point discussed in the *Notes* is the issue of singularities of the solutions of the field equations, at the core of the EIH argument mentioned above. A single particle at rest is, according to this argument, represented by a gravitational field that is finite and regular everywhere, except at the location of the particle. Einstein, Infeld, and Hoffmann, like Einstein and Grommer before them, used this special role of singularities in order to extract from the field equations the equation of motion by requesting that the gravitational field corresponding to the motion of two material particles is also nowhere singular, except at their position. As it turns out, the motions for which this is the case reduce in first approximation to the motions described by Newton's laws. Einstein saw this request to eliminate singularities outside of the positions of particles as a hint that a more satisfactory theory should avoid singularities altogether. He speculated that the solutions of the total field equations would describe the particles in other ways than by singular points and that the solution of equations of the total field would be free of singularities. He concludes the discussion of this point: "If one had the field equations of the total field, one would be compelled to demand that the particles themselves could be represented as solutions of the complete field equations that are free of irregularities everywhere. Only then would a general theory of gravity be a *complete* theory" (*Notes*, p. 77 [p. 179]). The fact that the singularity argument, originally going back to Einstein and Grommer, and now generally known as the EIH argument, pointed to this conclusion made it into the turning point to which we referred earlier.

At about the time of writing the *Notes*, Einstein presented a summary of his non-symmetric field approach to the goal of unification in appendix 2, "Generalized Theory of Gravitation," added to the third Princeton University Press (PUP) edition of his *The Meaning of Relativity* in 1950. He then significantly expanded his presentation and published it as appendix 2, under the same title, in the 1951 Methuen edition of that book. A modified version of this appendix was added to the fourth PUP edition in 1953 under the title "Generalization of Gravitation Theory." Einstein again significantly reformed this appendix for the fifth PUP edition, which appeared after his death, in 1956. In addition to the published versions, there are a number of unpublished pages of partial modifications of those versions. The different versions of this appendix summarize his persevering struggle to meet the challenge of finding a unified mathematical formulation of the entire physical reality. They contain, together with complicated mathematical derivations, numerous paragraphs with epistemological remarks on the purpose and meaning of his approach. The nonsymmetric field seemed to him, at this stage of his life, the most natural approach to achieve this goal.

Let us quote from that appendix a paragraph that clearly demonstrates that Einstein had undertaken this painstaking effort to formulate a comprehensive unified field theory in the hope that it could be an alternative to the contemporary probabilistic interpretation of quantum mechanics, which, for him, was an unacceptable description of physical reality:

> I must, however, explain why I have gone to so much trouble to arrive at this result. The contemporary physicist cannot, without such an explanation, appreciate this; for he is convinced, as a result of the successes of the probability-based quantum mechanics, that one must abandon the goal of complete descriptions of real situations in a physical theory. I do not want to discuss here why I do not share this conviction.... There is further the conviction that one cannot keep side by side the concepts of fields and particles as elements of the physical description.... The field concept, however, seems to be inevitable, since it would be impossible to formulate general relativity without it. . . . For this reason I see in the present situation no possible way other than a pure field theory, which then however has before it the gigantic task of deriving the atomic character of energy.[8]

In the *Notes*, Einstein presents the basic ideas of the nonsymmetric field approach using professional mathematical terms and expressions. We shall not attempt to interpret these pages. We have limited ourselves to comments on the incentives and goals of this approach.

Einstein's warning that "[e]very reminiscence is colored by one's present state, hence by a deceptive point of view" (p. 3 [p. 157]) does not apply to this part of his *Autobiographical Notes*. These are not autobiographical reminiscences but rather an account of his contemporary struggle. During the months of writing the *Notes*, he was already deeply involved in pursuing his nonsymmetric field approach, as has been demonstrated by his correspondence with Schrödinger. At that time, he was working with his assistant Ernst Gabor Straus on this and similar issues. A sheet of working paper in Einstein's handwriting, which came to us from the estate of Straus, shows at the top exactly the same equation used in his presentation of the subject in the *Notes* (eq. A, p. 93 [p. 182]).[9]

One of Einstein's "working pages" from the Straus estate. The first line is identical to equation (A) in the *Autobiographical Notes*. © Hebrew University of Jerusalem.

Einstein wrote supplementary remarks to the discussion of his nonsymmetric field approach, which he intended to include in the second edition of *Albert Einstein: Philosopher-Scientist*. These unpublished comments consist of two parts.[10] The first part is related to the mathematical formalism. Einstein points out that the field equations in this approach are not as uniquely determined by basic theoretical requirements as in the case of the symmetric field, confined to gravitation alone. He is not convinced that the choice presented in the *Notes* is the most natural one. After a brief discussion of the mathematical consequences of his choice of the equations and its inherent difficulties, Einstein shifts abruptly to epistemological remarks on the fundamental concepts underlying a comprehensive description of the world (see box). Possibly, these comments, which focus on the freedom of conceptual constructions in science, were triggered by his realization that his proposal for an extension of his gravitation theory was less constrained by his basic assumptions than he had initially assumed. These remarks could also serve as a supplementary expansion of Einstein's discussion of the process on thinking (*Notes*, pp. 7–11 [pp. 158–159]).

UNPUBLISHED SUPPLEMENTARY REMARKS TO THE *NOTES*

Everything that is conceptual is constructive and cannot be logically derived directly from experience. Thus, we have complete freedom in choosing the fundamental concepts on which we base our representation of the world. It all depends on how well suited our construction is for bringing order into the chaos of our world of experience.

Natural science, through a long process of development, has been brought to the stage of attempting to reduce everything to basic space-time concepts, which follow from the concept of a material object. In this sense, it is essentially "materialistic." Concepts coming from the psychological realm, such as will, personality, and so forth, are excluded as fundamental concepts, after it [natural science] has convinced itself through long struggles that the combination of fundamental concepts from the two spheres of concepts would not be fruitful.

In contrast to this they seek to reduce everything to basic concepts that stem from the psychological sphere (animism). To me it seems that all such conceptual systems do not achieve anything for the comprehension of the relations of the "external" experiences, and this is the case not only when considered from the vulgar utilitarian point of view.

As different as our strivings may be, they still have <u>one</u> principle in common; the assumption of a "real world" that separates, so to say, the "world" from the thinking and perceiving subject. The extreme positivists believe that they can renounce this as well; but this seems an illusion to me, if they are not prepared to renounce thinking all together.[11]

Einstein concludes the exposition of his nonsymmetric field program with the expectation: "I believe that these equations constitute the most natural generalization of the equations of gravitation." Actually, in the manuscript he wrote: "I am convinced that . . .", crossed it out, and replaced it with "I believe that . . .". The footnote attached to this sentence, which is not in the manuscript and must have been added in proof, may explain this change of words: "The theory here proposed, in my view, represents a fair probability of being found valid, if the way to an exhaustive description of physical reality on the basis of the continuum turns out to be possible at all." Contrary to most contemporary physicists, Einstein did not give up hope, until the end of his life, that it was indeed possible.

The last sentence of the *Autobiographical Notes* is a powerful summary of the entire effort and of its goal: "This exposition has fulfilled its purpose if it shows the reader how the efforts of a life hang together and why they have led to expectations of a certain kind" (*Notes*, p. 89 [p. 183]). He has just outlined his expectation about the ultimate nature of a complete physical theory, and he now indicates that his entire life effort led him to this expectation. This may be read as an echo of Einstein's formulation of the purpose of this effort, presented at the very beginning: ". . . I do, in fact, believe that it is a good thing to show those who are striving alongside of us how our own striving and searching appears in retrospect" (p. 3 [p. 157]).

The last page of the handwritten version of Einstein's *Autobiographical Notes*. Differences from the published version are indicated in the text. The Morgan Library & Museum, New York.

In the forty-five pages of the *Notes*, not divided into chapters and sections, Einstein presented his intellectual life as a flow of ideas and struggles, supporting each other and following from each other, all of them embedded in a consistent epistemological credo and a coherent scientific worldview. The unified field theory would represent the capstone of his scientific worldview and the epitome of his life effort. Einstein therefore saw the coherence of both, his worldview and his lifelong struggles, as hinging on the success of this effort. This becomes clear from a letter to his longtime friend Maurice Solovine: "You may imagine that I look back upon my life's work with silent contentment. But it is very different when seen from a personal point of view. There is not one single concept about which I am convinced it will last and I am uncertain as to whether I am even on the right track. The current generation sees in me both a heretic and a reactionary who has so to speak outlived himself."[12]

NOTES

1. Einstein, "Foundation of General Relativity," CPAE vol. 6, Doc. 30, p. 188.
2. Einstein, "Fundamental Ideas and Problems of the Theory of Relativity," in *Les Prix Nobel en 1921–1922*, ed. C. G. Santesson (Stockholm: Norstedt & Fils, 1923). English translation adapted from CPAE vol. 14, Doc. 75, p. 80.
3. Albert Einstein and Jakob Grommer. "Allgemeine Relativitätstheorie und Bewegungsgesetz," *Sitzungsber. phys-math. Kl.* 1 (1927): 235–245; Albert Einstein, Leopold Infeld, and Banesh Hoffmann, "The Gravitational Equations and the Problem of Motion," *Annals of Mathematics* 39 (1938): 65–100. For historical discussion, see Dennis Lehmkuhl, "General Relativity as a Hybrid Theory: The Genesis of Einstein's Work on the Problem of Motion," *Studies in History and Philosophy of Science Part B: Studies in History and Philosophy of Modern Physics*, 67 (2019): 176–190.
4. For a detailed survey of different approaches adopted by Einstein toward this goal, see Tilman Sauer, "Einstein's Unified Field Theory Program," in *The Cambridge Companion to Einstein*, ed. Michel Janssen and Christoph Lehner (Cambridge: Cambridge University Press, 2014). See also Jeroen van Dongen, *Einstein's Unification* (Cambridge: Cambridge University Press, 2010).
5. Einstein to Schrödinger, 22 January 1946, AEA 22-93.
6. Einstein to Schrödinger, 22 February 1946, AEA 22-98.
7. Einstein to Schrödinger, 20 May 1946, AEA 22-106.
8. Cited and discussed in Hanoch Gutfreund and Jürgen Renn, *The Formative Years of Relativity: The History and Meaning of Einstein's Princeton Lectures* (Princeton, NJ: Princeton University Press, 2017), 137.
9. We are grateful to Tilman Sauer for pointing us to this source.
10. Einstein, unpublished supplementary remarks, AEA 2-024.
11. The epistemological part of the supplementary remarks to pages 92–93 of the *Autobiographical Notes* (unpublished), AEA 2-024.
12. Einstein to Solovine, 28 March 1949, AEA 21-260.

PART III

EINSTEIN AND HIS CRITICS

1

THE PHYSICISTS AND PHILOSOPHERS WHO CONTRIBUTED TO THE VOLUME

IN HIS INTRODUCTORY REMARKS TO *ALBERT EINSTEIN: PHILOSOPHER-SCIENTIST*, ITS editor, Paul Arthur Schilpp, asserts that Einstein's *Autobiographical Notes* in itself assures the unique importance of the volume. He believed that the world would not possess the one and only intellectual autobiography of the illustrious scientist if the special nature of The Library of Living Philosophers (LLP) had not convinced Einstein to produce such an "obituary," as he calls his autobiography. In line with the structure of all the books in the series, the volume on Einstein also features descriptive and critical essays by twenty-five contributors.

The first of these essays, by Arnold Sommerfeld, was not written specifically for the Einstein volume. It was published elsewhere as a tribute to Einstein on his seventieth birthday and reprinted in the LLP volume. Schilpp planned to issue the book on 14 March 1949, the day of Einstein's birthday, but to his regret there was a slight delay in the publication.

Years later, describing the formative years of the LLP project, Schilpp recalled that Einstein, more than the other philosophers in this series, refused emphatically to have anything to do with the selection of contributors to his volume. He insisted that if the subject-philosopher participated in such a selection, it would violate the purpose of this series, namely, a free discussion between a philosopher and his critics and disciples.[1]

Among the essays of the contributors, Schilpp specifically emphasizes the significance of Niels Bohr's account of his conversations and debates with Einstein on the epistemological aspects of physics, which came into being only owing to the peculiar nature of this series. Schilpp regrets, to the point of considering it a tragedy, that Max Planck was too seriously ill to be able to contribute an essay. He also regrets that Hermann Weyl, one of the most active mathematicians-physicists during the formative years of general relativity, was unable to fulfill his promise to write an essay on general relativity and motion. Schilpp mentions, without specifying their names, three other scholars who failed to fulfill their pledges to the editor. From Schilpp's correspondence and earlier prospective lists of contributors, we know that two of them were the German mathematician Paul Epstein and the Russian physicist Yakov Frenkel.

Schilpp's initial list of prospective contributors to the Einstein volume contained a number of prominent physicists and philosophers who, giving various reasons, rejected the invitation to participate.[2] Among them we find the physicists Hans Bethe, Paul Dirac, Richard Feynman, Julian Schwinger, Erwin Schrödinger, the Indian-American astrophysicist Subrahmanyan Chandrasekhar, the Swiss-American astronomer Fritz Zwicky, and Einstein's assistant in Princeton, Valentine Bargmann. The philosophers who did not accept Schilpp's invitation are Bertrand Russell and Sir Edmund Whittaker. Schilpp tried several times, directly and with help of colleagues, to contact the Russian physicist Lev Landau, without any response whatsoever. It is not clear if his letters of invitation ever reached Landau.

In some cases, Schilpp corresponded with confirmed contributors on the subject and nature of their prospective essays and also in order to remind them of the deadlines for submission of their articles. For example, when approached, Wolfgang Pauli immediately agreed to contribute an essay to the Einstein volume. He did not approve the initially proposed title, "Towards a Merger of Quantum and Relativity Theory"; he preferred "Einstein's Contributions to Quantum Theory." However, this was the title of the contribution already promised by Victor Lenzen. Pauli thought that Lenzen, primarily a philosopher and not a physicist, was less qualified to write on this topic. Schilpp was not concerned by a possible duplication of titles. He argued that Lenzen had a position in the physics department at the State University of California, and he had no doubt that if the two of them happened to submit contributions under the same title, they would still be very different from each other. Finally, Lenzen changed his title to "Einstein's Theory of Knowledge." The correspondence with Pauli took place when the latter was planning to go to Stockholm to receive his Nobel Prize. After that he intended to visit Niels Bohr in Copenhagen. Pauli knew even before Schilpp that Bohr had agreed to contribute to the Einstein volume an essay on his discussions with Einstein and that he took this assignment very seriously. Pauli planned to coordinate his own contribution with Bohr.

There is also a long correspondence between Schilpp and the mathematician Kurt Friedrich Gödel. Schilpp expected to get from Gödel a comprehensive article that could be titled "The Realistic Standpoint in Physics and Mathematics." Gödel did not want to commit to a long article and agreed to contribute a short essay of three to five pages, if that would fit into Schilpp's scheme. He suggested to write "Some Remarks about the Relation between the Theory of Relativity and Kant." Schilpp would rather have a short paper than nothing at all. He did not like the "some remarks . . ." in the title, hoping that once Gödel began his writing the essay would expand to more than that. Three titles were mentioned: "The Philosophical Significance of Relativity Theory," "The Relation between the Theory of Relativity and Kant," and simply "The Theory of Relativity and Kant." Schilpp preferred the latter. This topic was on Gödel's mind. We know that at that time he received and studied an article by Ilse Rosenthal-Schneider (a contributor to the Einstein volume) on the theory of relativity and Kant.[3] Finally, Gödel submitted a relatively short article titled "A Remark about the Relationship between Relativity Theory and Idealistic Philosophy."

A list of the authors whose articles were finally included, with the title of their contributions and brief biographical notes, is presented below, in the order of their appearance in the book. The notes show why Schilpp invited these physicists and philosophers to contribute to the volume. We do not intend to present and analyze their remarks, beyond what will be implied by Einstein's response to them.

Arnold Sommerfeld (1868–1951): "To Albert Einstein's Seventieth Birthday" (not mentioned in Einstein's "Reply")

A German theoretical physicist, who served as professor at the University of Munich from 1906. He studied mathematics and later shifted to mathematical physics. Sommerfeld became one of the most important pioneers of atomic and quantum physics, about which he published a first fundamental book in 1919. He was an excellent academic teacher and formed an influential school of theoretical physicists. Because of his mastery of mathematical instruments, Sommerfeld was able to apply Einstein's special theory of relativity to different problems of physics and thus contributed to the establishment of the theory in the years 1907–1910. In this period, Sommerfeld and Einstein also met personally and discussed problems of the early quantum theory. Their scientific exchange continued in later years through intensive correspondence.

Louis de Broglie (1892–1987): "A General Survey of the Scientific Work of Albert Einstein" (not mentioned in Einstein's "Reply")

A French physicist who postulated in his PhD dissertation in 1924 that electrons and all particles have wave properties. This was demonstrated in 1927 by the way streams of electrons were diffracted by crystals. For his discovery of this wave-particle duality, which forms a central part of quantum mechanics, he won the Nobel Prize in 1929. In his later career, de Broglie developed a causal explanation of wave mechanics, in opposition to the generally accepted probabilistic interpretation. This was refined by David Bohm in the 1950s and has since been known as the de Broglie–Bohm theory. In addition to his scientific work, de Broglie thought and wrote about the philosophy of science. He was elected to the French Academy of Sciences and served as its perpetual secretary.

Ilse Rosenthal-Schneider (1891–1990): "Presuppositions and Anticipations in Einstein's Physics" (not mentioned in Einstein's "Reply")

A German-Australian physicist and philosopher. She earned her PhD in 1920 from the University of Berlin, where she first met Einstein. Her dissertation was on the problem of space-time in Kant and Einstein. In 1938, she had to leave Germany and settled in Australia, where she became a tutor in the German department at the University of Sydney in 1945 and taught history and philosophy of science. In the 1940s and 1950s, she exchanged a series of letters with Albert Einstein about philosophical aspects of physics, such as the theory of relativity, fundamental constants, and physical reality. She remained in contact with Einstein through correspondence until his death in 1955. In her later years, she published the book *Reality and Scientific Truth: Discussions with Einstein, von Laue, and Planck* (1980).

Wolfgang E. Pauli (1900–1958): "Einstein's Contributions to Quantum Theory"

After receiving his doctoral degree in 1921 as a student of Sommerfeld at the University of Munich, Pauli spent time at the universities of Göttingen, Copenhagen, and Hamburg before his appointment as professor of theoretical physics at the Federal Institute of Technology in Zurich. Pauli made groundbreaking contributions to modern physics, primarily in quantum mechanics. In 1945, he received the Nobel Prize in Physics for his discovery of a new law of nature, the exclusion or Pauli principle. The discovery involved a relation between the spin of the electron and the atomic structure, which is the basis of the theory of matter. Pauli was the first to recognize the existence of the neutrino, an uncharged particle of very small mass, which carries off energy in the radioactive decay of an atomic nucleus. He was one of the pioneers of the quantum theory of fields, and he participated actively in the great advances made in this domain during its fostering years. In 1921, when he was still a student, he published a comprehensive review of the theory of relativity, which became a seminal text during the formative years of general relativity. Contrary to Einstein, Pauli did not believe, even in the early stages of the quest for a unified field theory, that such a theory, which would also account for the origin and nature of elementary particles, is possible in the framework of classical continuous fields.

Max Born (1882–1970): "Einstein's Statistical Theories"

Max Born was one of the founders of modern quantum physics. He studied mathematics and physics at several universities. Among his teachers in Göttingen was Hermann Minkowski, who introduced him to electrodynamics and special relativity theory. Born's first publications, written between 1909 and 1914, were dedicated to electron theory, relativity theory, crystal physics, and Einstein's quantum theory of specific heat. In the following years, he focused his work on atom physics and the mathematical development of quantum physics. In 1915, Born was appointed professor of theoretical physics at the University of Berlin, where he became a close friend of Einstein. Their friendship is depicted in the classic collection the *Born-Einstein Letters, 1916–1955*. Later, he also taught in Frankfurt and from 1921 in Göttingen, where he formed a group that in 1925 formulated the foundations of quantum mechanics. In 1933, Born was forced to emigrate and went to Great Britain.

Walter Heitler (1904–1981): "The Departure from Classical Thought in Modern Physics"

A German physicist, who received his doctorate in theoretical physics in 1924 at the University of Munich. He then studied as a postdoctoral fellow with Niels Bohr at the University of Copenhagen and with Erwin Schrödinger at the University of Zurich. In Zurich, in 1927, Heitler applied the new quantum mechanics to deal with the problem of chemical valence bonding. This was a landmark in bringing chemistry under the realm of quantum mechanics. In 1933, when the Nazis came to power, Heitler began his work with Max Born at the University of Göttingen. Because of his Jewish origin, he had to leave Germany and became a research fellow at the University of Bristol, with Nevil Mott. Though the application of quantum mechanics to chemistry remained a prominent theme in Heitler's career, in Bristol he worked on quantum field theory, quantum electrodynamics, and the theory of cosmic rays. In 1936, he published his fundamental book, *The Quantum Theory of Radiation*. In 1941, Heitler moved to the Dublin Institute for Advanced Studies, where, in 1946, he succeeded Erwin Schrödinger as the director of the School for Theoretical Physics. In the 1960s and 1970s, back in Zurich, he turned his attention to the study of the relation between man, natural science, and religion.

Niels Bohr (1885–1962): "Discussion with Einstein on Epistemological Problems in Atomic Physics"

A Danish physicist who made fundamental contributions to understanding atomic structure and quantum theory, for which he received the Nobel Prize in Physics in 1922. He founded the Institute of Theoretical Physics at the University of Copenhagen, now known as the Niels Bohr Institute, which opened in 1920. He is one of the founders and promoters of the so-called Copenhagen interpretation of quantum mechanics. After World War II, Bohr was active in the international arena, calling for cooperation on nuclear energy. He participated in the establishment of CERN (European Organization for Nuclear Research) and of the Research Establishment Risø of the Danish Atomic Energy Commission, and became the first chairman of the Nordic Institute for Theoretical Physics in 1957. During the 1920s and the 1930s, Bohr and Einstein expressed their different opinions on quantum mechanics in a number of debates. These debates represent one of the high points in physical research in the first half of the twentieth century. They are also remembered because of their importance to the philosophy of science. Bohr's contribution to the volume on Einstein in The Library of Living Philosophers is a comprehensive account of these debates. Despite the disagreements, their friendship and mutual respect lasted throughout their lives.

Henry Margenau (1901–1997): "Einstein's Conception of Reality"

A German-US physicist, and philosopher of science. He wrote extensively on science, and his works include: *Ethics and Science*, *The Nature of Physical Reality*, *Quantum Mechanics*, and *Integrative Principles of Modern Thought*. Margenau received a PhD degree from Yale University in 1929. From 1950 until his retirement from formal academic life in 1986, he served as a professor of physics and natural philosophy at Yale University. He also acted as staff member at the Institute for Advanced Study, Princeton University, and the MIT Radiation Laboratory. He adopted indeterminism as a first step toward a model of free will. Margenau served on a commission of the World Council of Churches in developing an ecumenical position on nuclear weapons and atomic warfare. His book *The Miracle of Existence* shows his interest also in Eastern religions and in connections among different religious and philosophical traditions.

Philipp G. Frank (1884–1966): "Einstein, Mach, and Logical Positivism" (not mentioned in Einstein's "Reply")

An Austrian, and later, an American mathematician and philosopher during the first half of the twentieth century. He studied physics at the University of Vienna and graduated in 1907 with a thesis in theoretical physics under the supervision of Ludwig Boltzmann.

Albert Einstein recommended him as his successor at the German Charles-Ferdinand University of Prague, a position that he held from 1912 until 1938. He was a logical positivist, influenced by Mach, and a member of the Vienna Circle. He published a biography, *Einstein: His Life and Times*. In 1938, he emigrated to the United States, where he became a lecturer in physics and mathematics at Harvard University. In 1947, he founded the Institute for the Unity of Science as part of the American Academy of Arts and Sciences. The institute held regular meetings that attracted a broad range of participants and was regarded as the "Vienna Circle in exile."

Hans Reichenbach (1891–1953): "The Philosophical Significance of the Theory of Relativity"

Reichenbach was born in Hamburg. He studied physics, mathematics, and philosophy in Berlin, Göttingen, and Munich. He received his PhD in Erlangen with a dissertation on the concept of probability. In the winter semester of 1917/18 he continued his studies in Berlin, where he took a course on relativity with Einstein. Together with Moritz Schlick he soon became one of the most prominent advocates and interpreters of relativity theory. He criticized popular misunderstandings of the theory and developed his own views on science in the context of philosophical discussions on relativity. He was one of the founders of logical positivism. Following a proposal by Einstein, in 1926 Reichenbach became professor of the philosophy of physics at the University of Berlin. He was dismissed in 1933, after the rise of the Nazi regime, and emigrated to Turkey, where he became professor at the University of Istanbul. In 1938, he emigrated to the United States, where he taught, until the end of his life, at the University of California, Los Angeles.

Howard P. Robertson (1903–1961): "Geometry as a Branch of Physics"

An American mathematician and physicist, served as professor of mathematical physics at the California Institute of Technology and Princeton University. Robertson was one of the pioneers of the formative years of relativistic cosmology and the emergence of the paradigm of an expanding universe. His name is associated with the Poynting–Robertson effect, the process by which solar radiation causes a dust particle orbiting a star to lose angular momentum, describing this effect in terms of general relativity. His best-known achievements were in applying the general theory of relativity to cosmology.

Percy Williams Bridgman (1882–1961): "Einstein's Theories and the Operational Point of View"

An American physicist who won the 1946 Nobel Prize in Physics for his work on the physics of high pressures. He also wrote extensively on the scientific method and on other aspects of the philosophy of science. His book *The Logic of Modern Physics* (1927) advocates operationalism and coined the term "operational definition." In 1938, he participated in the International Committee composed to organize the International Congresses for the Unity of Science. He was also one of the eleven signatories to the Russell–Einstein Manifesto.

Victor F. Lenzen (1890–1975): "Einstein's Theory of Knowledge"

An American physicist and philosopher of science. Lenzen began his undergraduate studies in physics at the University of California, shifted to philosophy, and received a PhD in philosophy from Harvard University in 1916, studying with Bertrand Russell and Josiah Royce, whose seminar on scientific methodology influenced him profoundly. Apparently, he did not agree with Royce's idealism, in the sense that he was more interested in physical models and concepts, and he returned to physics as a profession. After short periods in Cambridge, England, and at Harvard, he began his life career in Berkeley, California. His major book, *The Nature of Physical Theory*, published in 1931, is devoted to a critical analysis of the concepts, principles, and systems of a physical theory. He expanded his views on the metaphysical and methodological questions posed by physical theory.

Filmer S. C. Northrop (1893–1992): "Einstein's Conception of Science"

An American philosopher. His most influential work, *The Meeting of East and West*, was published in 1946 at the aftermath of World War II. Its central thesis is that East and West both must learn something from each other to avoid future conflict and to flourish together. He was the author of twelve books and innumerable articles on all major branches of philosophy including epistemology and the theory of concepts.

Edward A. Milne (1896–1950): "Gravitation without General Relativity"

An American theoretical astrophysicist. Milne's early work was in mathematical astrophysics. In the 1920s, much of his research was concerned with stars, particularly with stellar evolution. His research in the 1930s was mainly concerned with the theory of relativity and cosmology. His later work, concerned with the interior structure of stars, aroused controversy. Milne was president of the Royal Astronomical Society from 1943 to 1945. In his work *Relativity, Gravitation, and World-Structure* (1935), he proposed an alternative to Einstein's general relativity theory, deriving a cosmological model of an expanding universe with an inhomogeneous mass distribution within the special theory of relativity.

Georges Lemaître (1894–1966): "The Cosmological Constant"

Alongside theological studies, the Belgian catholic priest Georges Lemaître pursued research in astrophysics, cosmology, and mathematics. In 1927, he was appointed professor of physics at the University of Louvain. As early as 1925, Lemaître worked on the application of Einstein's general theory of relativity to cosmology, and in 1927 published a fundamental paper in which he solved the equations of the gravitational field without using Einstein's cosmological constant. He developed solutions of general relativity for an expanding universe, as Alexander Friedmann had done some years earlier. Lemaître delivered a demonstration of Hubble's law of recessional velocity before Hubble himself. He was the first to propose an early version of the big-bang theory, suggesting that the universe evolved from a primeval atom.

Karl Menger (1902–1985): "Modern Geometry and the Theory of Relativity"

An Austrian-American mathematician. He received his PhD from the University of Vienna in 1924. He was an active participant of the Vienna Circle, which in the 1920s also discussed social science and philosophy. During that time, Menger proved an important result on the so-called St. Petersburg paradox with interesting applications to the utility theory in economics. Later he contributed to the development of game theory with Oskar Morgenstern. He taught at the University of Amsterdam, University of Vienna, University of Notre Dame, and Harvard University, but his longest and last academic post was at the Illinois Institute of Technology. In mathematics, he worked on the mathematics of algebras, algebra of geometries, curve and dimension theory, and on many other topics. He is considered one of the founders of distance geometry, by having formalized definitions of the notions of "angle" and of "curvature" in terms of directly measurable physical quantities.

Leopold Infeld (1898–1968): "General Relativity and the Structure of Our Universe"

A Polish theoretical physicist who received his doctoral degree from the Jagellonian University in Cracow in 1921. During the period 1930–1938, he worked with Einstein at Princeton on the problem of motion in the general theory of relativity. In 1938, he coauthored with Einstein the popular book *The Evolution of Physics*, which generated a lot of public interest and became a bestseller. Between 1939 and 1950, he served as a professor at the University of Toronto, where he conducted pioneering research on magnetism in general relativity. After the first use of nuclear weapons in 1945, Infeld, like Einstein, became a peace activist. Because of his activities, he was unjustly accused of having communist sympathies. In 1950, he left Canada and returned to communist Poland to become a professor at the University of Warsaw, a post he held until his death. He felt he had an obligation to contribute to the recovery of science in Poland after the ravages of World War II. He was stripped of his Canadian citizenship and was widely denounced as a traitor. After Infeld's return to Poland, he requested a leave of absence from the University of Toronto. His request was denied, and he resigned his post. In 1995, the University of Toronto granted Infeld the posthumous title of professor emeritus. Infeld was one of the eleven signatories to the Russell–Einstein Manifesto in 1955, and he is the only signatory never to receive a Nobel Prize.

Max von Laue (1879–1960): "Inertia and Energy"

The German theoretical physicist Max von Laue was appointed as lecturer at the University of Berlin in 1906. From 1914 to 1919, he was professor at the University of Frankfurt, and from 1919, at the University of Berlin. Laue was particularly concerned with mathematical aspects of problems of optics and in 1907 delivered a mathematical explanation of a problem of light propagation in the frame of Einstein's special theory of relativity. Laue's work fostered the acceptance of the theory. He also contributed to its development, in particular with his work on relativistic continuum dynamics and published, besides many articles, a first book on the subject in 1910 and a second one in 1919. In addition, Laue concerned himself with the theory of X-ray interference with matter and with the theory of superconductivity. In 1912, he discovered the diffraction of X-rays by crystals. After their first meeting in 1906, Laue and Einstein became lifelong friends.

Herbert Dingle (1890–1978): "Scientific and Philosophical Implications of the Special Theory of Relativity"

An English physicist and natural philosopher who served as president of the Royal Astronomical Society from 1951 to 1953. He is best known for his opposition to Einstein's special theory of relativity and the protracted controversy that this provoked. His initial campaign to investigate the respective merits of Henri Bergson and Einstein soon became a broad controversy about social planning and the value of science education over one based on the arts and humanities. Dingle gained the reputation of a stubborn crackpot who refused to accept the achievements of Einstein.[4]

Kurt Friedrich Gödel (1906–1978): "A Remark about the Relationship between Relativity Theory and Idealistic Philosophy"

An Austrian, and later an American logician, mathematician, and philosopher who made a great impact on scientific and philosophical thinking in the twentieth century. Gödel received his PhD from the University of Vienna in 1930, and shortly after that, at the age of twenty-five, he published his incompleteness theorems, which brought him international fame. After the rise of the Nazis to power in Germany and the assassination of one of his mentors, Moritz Schlick, in 1936, he developed paranoid symptoms and spent several months in a sanitarium for nervous diseases. In 1940, he joined the Institute for Advanced Studies in Princeton, where he had lectured on several occasions during the 1930s. Einstein was also living at Princeton at that time, and they developed a close friendship. During his many years at the institute, Gödel's interests turned to philosophy and physics. In 1949, he found a new exact solution of Einstein's field equations of general relativity. This solution has unique properties—it models a rotating universe, and it permits the existence of closed timelike curves, which would allow travel in time. Apparently, he gave this result to Einstein as a present on his seventieth birthday.

Gaston Bachelard (1884–1962): "The Philosophic Dialectic of the Concepts of Relativity" (not mentioned in Einstein's "Reply")

A French philosopher who contributed to the fields of poetics and the philosophy of science. In the latter, he introduced the concepts of "epistemological obstacle" and "epistemological break." For him, scientific developments, such as Einstein's theory of relativity, demonstrated the discontinuous evolution of science. Thus, models that framed scientific development as continuous seemed to him simplistic and erroneous. He argued that new theories integrated old theories in new paradigms, changing the sense of concepts (for instance, the concept of mass, used by Newton and Einstein in two different senses). In his view, the role of epistemology is to explore the history of the production of scientific concepts.

Aloys Wenzl (1887–1967): "Einstein's Theory of Relativity Viewed from the Standpoint of Critical Realism, and Its Significance for Philosophy" (not mentioned in Einstein's "Reply")

A German philosopher who studied physics and mathematics at the University of Munich in 1912, where he taught philosophy and psychology (1926–1938). He was also interested in parapsychology. He was then discharged from teaching on ideological grounds during by the Nazi regime. Wenzl returned to his position as professor of philosophy at the University of Munich in 1946 and served as rector of the university in 1947–1948. In 1924, he published his work on the relation between Einstein's theory of relativity and contemporary philosophy with particular emphasis on the philosophy of "as if," which was praised by Albert Einstein, Max von Laue, and Moritz Schlick. The philosophy of "as if," formulated by Hans Vaihinger, argues that sensations and feelings are real, but the rest of human knowledge consists of "fictions" that can only be justified pragmatically. In his view, even the laws of logic are fictions, albeit fictions that have proved their indispensable worth in experience and are thus held to be undeniably true. Wenzl's best-known book is *Philosophy of Freedom* (1947).

Andrew Paul Ushenko (1900–1956): "Einstein's Influence on Contemporary Philosophy" (not mentioned in Einstein's "Reply")

Ushenko was born in Moscow and participated in the Bolshevik revolution. In 1925, he emigrated to the United States. He was accepted to the graduate program in mathematics at the University of California, Berkeley. He switched to philosophy and received his doctorate in philosophy in 1927. Ushenko taught philosophy at the University of Michigan, Princeton University, and Indiana University. He worked primarily on the philosophy of science, epistemology, metaphysics, and philosophy of logic. He was especially interested in the nature of change and time, the theory of relativity, the theory of meaning, and the conflict between classical logic and new methods in logic. Ushenko also contributed an essay to the LLP volume on Bertrand Russell.

Virgil Hinshaw (1920–1995): "Einstein's Social Philosophy" (not mentioned in Einstein's "Reply")

An American philosopher who received his PhD degree at Princeton and spent his entire professional career at Ohio State University. In graduate school he was influenced by Bertrand Russell and Albert Einstein, and published philosophical writings about both of them. He was acknowledged among post–World War II American philosophers, especially for his work on the theory of knowledge and philosophy of history, sociology, and natural sciences.

NOTES

1. Schilpp, "Glimpses of a Personal History," Special Collections Research Center, Southern Illinois University Carbondale, box 21, folder 2.
2. Schilpp's invitation letters and correspondence with prospective and confirmed contributors to the Einstein volume of LLP may be found in the Paul Arthur Schilpp papers, Special Collections Research Center, Southern Illinois University Carbondale, box 15, folders 6, 8, 9.
3. Einstein to I. Rosenthal-Schneider, 3 February 1947, AEA 20-281.
4. Jimena Canales, *The Physicist and the Philosopher: Einstein, Bergson, and the Debate That Changed Our Understanding of Time* (Princeton, NJ: Princeton University Press, 2015), pp. 189–193.

2
EINSTEIN'S
"REPLY TO CRITICISMS"

How can one adequately praise the care, precision, directness, and
beauty of Professor Einstein's "Reply" (or "Remarks," as he calls them)
to his commentators and critics!

—Paul A. Schilpp, introduction to *Albert Einstein:
Philosopher-Scientist*, p. xiv

THE FIRST TWO OF THE TWENTY-FIVE ESSAYS BY EINSTEIN'S CONTEMPORARIES, SCIEN-
tists and philosophers, are different from the others. The article by Arnold Sommerfeld
"To Albert Einstein's Seventieth Birthday" has already been mentioned. Similarly, the
article by Louis de Broglie, "A General Survey of the Scientific Work of Albert Einstein,"
is also a tribute to Einstein's work, an ode to his scientific achievements. Einstein did not
mention them in his "Reply to Criticisms."

Undertaking the task of responding to the other twenty-three essays, Einstein faced a
serious challenge. When he agreed to furnish a reply to the critical essays in the volume,
he made one reservation: "It is, however, understood that, in my 'Reply,' I shall enter into
a discussion of the criticisms of the contributors only as appears necessary and important
to me."[1] Initially, he intended to respond separately to each of the contributions that he
decided to address. He soon realized that, because of the large number of essays and the
variety of subjects and arguments, the result would be an inhomogeneous collection of
unrelated texts, which would neither give a coherent picture nor provide enjoyable reading. He therefore discarded the responses that he had already written. (The unpublished
responses are found in the Albert Einstein Archives; in our book they are referred to
by their archival call number). Instead, he composed a single "Reply" organized around
groups of essays and according to thematic guidelines. The result is a fluent account of
Einstein's views on contemporary physics, specifically on the generally accepted non-
deterministic character of quantum mechanics, and of the basic elements of his philosophical and epistemological thinking. It is a valuable supplement to his *Autobiographical
Notes*.

In the "Reply," Einstein focuses his remarks on a number of issues, referring specifically only to about one-half of the contributing authors. A few others are mentioned only briefly, and the rest are completely ignored—either because their contributions arrived too late or because, as he explains, their "mentality . . . differs so radically from my own, that I am incapable of saying anything useful about them."[2] Initially, Einstein had written replies to some of the articles that he eventually decided not to mention in the published version. One such example is his treatment of the article by Andrew P. Ushenko, "Einstein's Influence on Contemporary Philosophy." In unpublished remarks, Einstein states that he can say nothing about this article because of the many terms used, which are not defined sharply enough for the non-philosopher. He makes a critical remark about Ushenko's use of the term "metaphysics."[3] Not being sure that he understood the article, and being even less sure that his remarks will help the reader to understand it, Einstein omitted any reference to this article in the published reply.

When Ushenko heard about it, he wrote to Einstein expressing the concern that Einstein did not find his article worth mentioning because he disapproved of it in some way. If this was not the case, he requested that Einstein add a line to that effect in the "Reply."[4] Einstein answered Ushenko that "such omission should not be interpreted as a tacit expression of unfavorable opinion but rather as an expression of modesty due to the limitations of my intellectual faculties. You will agree with me that it is better to say nothing than to say something incompetent."[5] Einstein did not comply with Ushenko's request to add a line about his article in the "Reply." It was too late for that in September 1949. Actually, Schilpp himself was very critical of Ushenko's article and regretted that he invited him to contribute to this volume, and if he could, he would have rejected it.[6]

Likewise, the long article by Aloys Wenzl, "Einstein's Theory of Relativity Viewed from the Standpoint of Critical Realism, and Its Significance for Philosophy," is not mentioned in the "Reply," although there is a brief unpublished remark about the general role of philosophy, related to this article: "Philosophy seeks the clarification of concepts and of thinking. Philosophy also seeks to bring seemingly diverse ideas and insights under a unified viewpoint. But in pursuing the second goal, the first one should not be abandoned."[7] Einstein referred differently to the essay by Herbert Dingle, "Scientific and Philosophical Implications of the Special Theory of Relativity." The brief remarks on this article in the "Reply" begin with the statement that, in spite of his efforts, he did not understand either its essence or its aim. In the unpublished response, this statement is followed by about two pages of critical comments.[8] They are omitted in the published version and replaced by several questions reflecting Einstein's criticism.

Another, very brief, but this time very complimentary, remark in the "Reply" is on the contribution of Einstein's colleague and friend, the physicist Max von Laue, "Inertia and Energy." Einstein describes this article as a historical investigation of the conservation principles that is of lasting value and deserves to be published independently and be made available to students. In the unpublished remarks, Einstein adds to this statement a reservation about von Laue's treatment of the relation between mass and energy, which he did not include in the published version.[9]

We would like to comment on two more of the essays not mentioned in Einstein's "Reply." The essay by Virgil Hinshaw, "Einstein's Social Philosophy," is the only contribution not related to Einstein's science and epistemology. When Einstein received this article, he suggested to Schilpp that it be treated with caution.[10] The author did not know

Einstein, and Einstein had the impression that the contribution lacked intellectual rigor. Schilpp responded in a detailed letter and sent Hinshaw's article back to him for improvements.[11] He was still unhappy with the result and asked for Einstein's advice about how to proceed. Einstein did not want to interfere because he did not think that it would be appropriate for him to influence a decision concerning an article about himself. He did not even read it, because he did not like reading about himself.[12]

The other article is by Philipp Frank, "Einstein, Mach, and Logical Positivism." It was very appropriate to invite Frank to contribute to this volume. Einstein and Frank knew each other; Frank was an active member of the Vienna Circle; he had written a biography of Einstein, which appeared in English in 1947; and it was he who suggested to Schilpp to bring Einstein into his project of The Library of Living Philosophers. It would have been interesting to read Einstein's response to this article. In the concluding remarks in his "Reply," Einstein indicates that he did not respond to contributions that arrived after the end of January 1949. Frank's essay was delivered at about that time. As the publication of the volume was delayed, Einstein had enough time to respond, but he thought that he had already written enough and did not want to add to it, although he wrote to Schilpp that Frank's work was excellent.[13] At this stage, he was tired of the whole project. To a friend he wrote: "I am sweating the whole time over the answers to the articles in the volume edited by Dr. Schilpp. If I had realized in time what this means, I would not have given my consent. I write, and afterwards I do not like it, and then I start over again. To hell with it!"[14]

We shall now summarize Einstein's reply to the essays that are closer to his own thinking and that he chose to address in greater detail. We highlight the main points in his remarks, emphasizing their place and role in his scientific, philosophical, and epistemological worldview. Whenever appropriate, we also refer to and quote from the unpublished texts mentioned above.

A. RESPONSE TO MAX BORN, WOLFGANG PAULI, WALTER HEITLER, NIELS BOHR, AND HENRY MARGENAU

Einstein refers to these scientists as his highly esteemed colleagues, who are firmly convinced that quantum mechanics is the final and satisfactory theory explaining the wave-particle duality and that one can only make statistical statements about measurements performed on microscopic physical systems. At the beginning, Einstein mentions specifically the articles of Born and Pauli, praising their meritorious historical description of his contributions to physical statistics and quanta. Both regret the fact that Einstein rejects the basic idea of contemporary quantum mechanics. Responding to them and to the other colleagues, Einstein uses the opportunity to explain once more, as he did in the *Notes*, but in greater detail, why he does not fall in line with them. He is firmly convinced that the statistical nature of this theory is only due to its incomplete description of physical systems and that this approach cannot provide a useful basis for a comprehensive theory of physics.

Before explicating Einstein's arguments, we shall briefly comment on two unpublished responses, to Born and Heitler.[15] Einstein refers to Born as a beloved friend, who believed that God throws dice. Quoting from Einstein's obituary article on Ernst Mach,[16] Born concludes: "That is the core of young Einstein, thirty years ago. I am sure the principles of probability were for him of the same kind as all other concepts used for describing nature. . . . The Einstein of today is changed."[17] He then quotes from a letter he received from Einstein (7 September 1944): "In our scientific expectation we have grown antipodes. You believe in God playing dice and I in perfect laws in the world of things existing which I try to grasp in a wildly speculative way."[18]

In his unpublished remarks, Einstein specifically responds to Born's allusion that his own opinion has changed over the years: "Born is doing me wrong in one aspect: he thinks that I have changed because previously I was using statistical methods, but the truth is that I never believed that the fundament of physics consists of statistical contents. The reason for our diverse expectations about the future development of the fundaments of physics can be easily seen: he thinks like most physicists of the present and in contrary to me, that the only possible definition of Quantum Theory is the only valid one."[19]

In this unpublished text, Einstein also comments on Born's remark about Einstein's quest for ". . . a general field theory which preserves the rigid causality of classical physics and restricts probability to masking our ignorance of the initial conditions."[20] He admits that this was indeed his goal. He did not accept the dominant conviction about the inadequacy of a continuous field theory as a basis of a comprehensive physical theory. He discusses this point at length in the *Autobiographical Notes* (see Part II, chapter 13).

The unpublished response to Heitler's article "The Departure from Classical Thought in Modern Physics" summarizes briefly and precisely the two sides of the controversy.[21] The philosophical problem lies in the relation between the "wave function" ψ, which describes a physical system in quantum mechanics and the reality of that system in space and time. The common understanding of quantum mechanics is that this function represents a complete as possible description of the physical situation and that a sudden change in the ψ-function by a new observation corresponds to a change in the physical reality generated by the observation. An alternative interpretation would be that a change in ψ does not correspond to a change in the real physical situation but rather to a change in our knowledge about this situation. Einstein is not convinced by Heitler's argument about the completeness of the description by the ψ-function, because a similar argument could be used to claim that the phenomenological theories of heat conduction are complete, while we know that a molecular-kinetic description of this phenomenon is more precise.

In his published "Reply," Einstein acknowledges the important and unparalleled progress that the "statistical quantum theory" has brought to physics. His dissatisfaction with this theory is related to its failure to fulfill the basic aim of all of physics: "the complete description of any (individual) real situation (as it supposedly exists irrespective of any act of observation or substantiation)."[22] To demonstrate his position, Einstein now deploys an argument not used in the *Notes*. He discusses the case of a single radioactive atom with a definite average decay time, which is localized at a point in space. The radioactive process is the emission of a relatively light particle through a potential barrier surrounding the atom. This particle is described in quantum mechanics by the function ψ, which tells us, at a given time, the probability that the particle is in a certain region of space. At $t = 0$, this function is confined to the interior of the atom, and it then spreads out. At every point in time, we can derive the probability that the emission has already occurred, but it does not provide any information about the time of disintegration of the radioactive atom. Einstein asserts that the ψ-function is, therefore, not a complete description of the radioactive process of an individual atom.

Einstein then presents the presumed answer of his opponents, the theorists of statistical quantum mechanics, to this statement. They would argue that assigning a definite disintegration time to an individual atom is arbitrary and meaningless, because there is no way to determine that time empirically, without disturbing the atom. Whatever the result of such a determination of the time of disintegration would be, it would provide no information on the status of the undisturbed atom. The critics of Einstein would argue that the problem lies in his claim that something not observable is "real." Einstein objects to this argument. Restricting physical statements about what is real to what is directly observable seems to him naive and untenable. It represents an exaggerated expression of the basic attitude of logical positivism, the dominant philosophy of science at that time, according to which only statements that can be empirically validated are meaningful.[23]

Einstein's discussion of the line of thinking of theoreticians of quantum mechanics leads to the consequence that it is meaningless for them even to ask if a definite instant of time of the disintegration of a single atom exists. If such a theoretician would assign to the function ψ the meaning of referring to an ensemble of atoms and if the derived probability that radioactive decay has already occurred represents the average over the individual atoms of this ensemble, he could assume a definite time of the disintegration. If, however, he insists that the function ψ is a complete description of an individual atom, he must reject the notion of a specific decay time.

Einstein continues with a long discussion of a system composed of a Geiger counter and an advancing strip of paper, where each decay, measured by the Geiger counter, is marked by a dot on the strip of paper. The location of the dot on the strip of paper marks the time of disintegration. Now, the interpretation of this time instant is in the framework of macroscopic concepts, completely distinct from the notion of the instant of disintegration of a single atom. Because it is a macroscopic system, it is to be expected that the location of the dot can be determined with certainty.

Einstein concludes this part of his remarks with a concise formulation of his conviction on this issue: "I am convinced that everyone who will take the trouble to carry through such reflections conscientiously will find himself finally driven to this interpretation of quantum-theoretical description (the function ψ is to be understood as the description not of a single system but of an ensemble of systems)."[24] He is convinced that within the framework of future physics, quantum mechanics will take an approximately analogous position to that of statistical mechanics within the framework of classical mechanics, but the path will be long and difficult.

Einstein invokes again a possible reaction of a theoretician of quantum mechanics. Suppose that such a theoretician would accept the conclusion that the quantum-mechanical description is an incomplete description of an individual system. He could then claim that the search for a complete description would be aimless, because the laws of nature cannot be completely formulated within the framework of an incomplete description. This could be a theoretical possibility. Einstein's point of view, however, is that ". . . the expectation that the adequate formulation of the universal laws involves the use of *all* conceptual elements which are necessary for a complete description, is more natural."[25]

This virtual dialogue has brought Einstein back to fundamental epistemological issues. The thrust of the argument that follows is his denial that quantum theory forces upon us a specific epistemological position because of the role measurements play in it. At the end of a longer digression on epistemology in which he recapitulates his own position, Einstein takes issue with Bohr's principle of complementarity of which he himself has tried, without success, to give a sharper formulation. In the context of this discussion, this principle means that systems, at the microscopic level, have certain pairs of complementary properties that cannot be accurately measured at the same time. The typical complementary phenomenon is the particle wave duality of a material particle. The type of measurement determines which of these two properties of the system emerges. It seems to Einstein that the epistemological error of this principle is to introduce a theoretical description which depends directly on empirical statements. To this Einstein objects. For him, the relation between sense experiences and thinking leading to a theoretical description is much more indirect and allows for a much greater freedom in the choice of appropriate concepts than such a quantum epistemology would suggest.

It is in order to justify this fundamental objection to the Copenhagen interpretation of quantum mechanics that Einstein recapitulates his epistemological credo in the paragraphs preceding his challenge of Bohr's position. This recapitulation is, at the same time, directed against the positivist prohibition of metaphysics and the ban of all concepts not directly related to empirical evidence. Einstein's position involves setting up a number of conceptual assertions required to think physically and to acquire knowledge of reality. They precede experience and thus may be considered, in a Kantian sense, a priori, with the difference, however, that Einstein does not consider such categories as being unalterable but as free conventions. He still claims that they are a priori in the more general sense that "thinking without the positing of categories and of concepts in general would be as impossible as is breathing in a vacuum."[26]

This is for Einstein enough to justify his plea to be "guilty of the metaphysical 'original sin,'" that is, of having introduced concepts that themselves lack direct contact with empirical evidence. In the philosophical tradition, metaphysics deals with fundamental questions about the world and its contents, without being restricted to scientific notions in the modern sense. Similarly, in his comment on Russell's theory of knowledge, Einstein speaks of "a fateful fear of metaphysics," describing it as a "malady of contemporary empiristic philosophizing." He finds also in Russell's work some hints supporting his belief that ". . . one can, after all, not get along without 'metaphysics.' The only thing to which I take exception there is the bad intellectual conscience which shines through between the lines."[27] While Russell had a bad conscience, Einstein did not. Unfortunately, Bertrand Russell, although invited to contribute to the Einstein volume of the LLP, did not reciprocate Einstein's willingness to contribute to his volume.

At the end of this excursion, Einstein returns to the key issue of his dispute with statistical quantum theory, the concept of the real in physics, contested by the Copenhagen interpretation of quantum physics. As we have seen, Einstein makes a point of deriving his own epistemological position not from a physical theory but from more fundamental considerations shaped by neo-Kantian reflections on the possibility of experience. In line with his remarks in the *Notes*, Einstein follows Kant also in his claim to conceive the real "as a type of program." He points out that nobody would easily abandon this program in the macroscopic realm and that, in his view, there are hardly any motives for doing so in the microscopic realm as well, in particular given the interrelations between the microscopic and the macroscopic. His punch line is the accusation against the quantum theorists of themselves taking an a priori position, that of clinging "to the thesis that the description of nature by the statistical scheme of quantum-mechanics is final."[28]

In the conclusion of the response to his colleagues, the physicists, Einstein goes back to the essays by Born and Pauli (in the text it says Bohr, but it is clear from the context and from the manuscript that he meant Born). Both accuse him, in a friendly fashion, of "rigid adherence to classical theory." Looking back at the transition from Newtonian physics to Maxwell's theory, in which forces acting at a distance were replaced by continuous fields, followed by the field theory of gravitation, which is still not a theory that explains the existence of masses, it is not clear, however, what is the "classical theory" to which one can claim that he adheres. Yet the continuous four-dimensional field theory exists as a program, and Einstein admits rigid adherence to this program. He then explains the reasons for this commitment, as he did when he discussed his quest for a unified field theory in the *Autobiographical Notes* (see Part II, chapter 13). However, he was also open to other alternatives, such as a purely algebraic theory, as we shall see in his response to Margenau.

B. RESPONSE TO HANS REICHENBACH

Einstein begins the discussion of the relation between the theory of relativity and philosophy with a commentary on the article by Hans Reichenbach, "The Philosophical Significance of the Theory of Relativity," which is characterized, as he writes, by "the precision of deduction and by the sharpness of his assertions." In this context, he also mentions the article by Robertson, "Geometry as a Branch of Physics," which is marked by a "lucid discussion" and is interesting from the standpoint of general epistemology, although it is confined to the narrower topic of the theory of relativity and geometry.[29]

In an unpublished response to Reichenbach's essay, Einstein is even more laudatory: "What distinguishes Reichenbach among so many of his colleagues is the circumstance that he never buys the generality of insight by sacrificing clarity. He sees in the logical critique of the doctrines and methods of the single sciences the main task of philosophy."[30] Einstein then comments on two points. First is Reichenbach's assertion that "[t]he logical basis of the theory of relativity is the discovery that many statements, which were regarded as capable of demonstrable truth or falsity, are mere definitions."[31] Einstein cannot confirm this statement as it is formulated here. He assumes that what Reichenbach meant was that the theoretical clarification that led to the special and general theory of relativity required an examination of basic concepts of geometry, in order to recognize what was merely of conventional nature. This resulted in a more useful scope of concepts. The analysis of concepts was in both cases a necessary tool but not the starting point (not a logical basis).

The second point on which Einstein commented is Reichenbach's definition of congruence: "That a certain distance is congruent [i.e., coinciding when superimposed] to another distance situated at a different place, can never be proven to be true.... [I]t can be maintained as true only after a definition of congruence is given; it therefore depends on an original comparison of distances which is a matter of definition."[32] Einstein comments that this delicate point stems from the fact that comparison of distances depends on the transport of a rigid body, which serves as a measuring device but constitutes, at the same time, a somewhat problematic concept. He addresses this ambiguity in his subsequent explanation: "The delicate point concerning this psychologically almost unavoidable opinion, lies in the fact that the inflexible measuring body presents a fiction whose justification is not beyond doubt. When trying to avoid this fiction, a direct physical definition of the geometric congruence is no longer possible. In that case, the thesis 'the meaning of a statement is reducible to its verifiability' appears to be in general problematic. It is in fact doubtful whether one can stick to this conception of 'meaning' for a single statement."

Einstein discusses here the alternative between a naive realism according to which geometrical distances can be established by using an inflexible measuring body, which he characterizes as psychologically almost unavoidable, and the claim that the comparison of distances is merely a matter of definition. In the latter case, the problematic character of the assertion that one can assign meaning to a single statement by reducing it to its verifiability becomes particularly evident.

In the published response, Einstein takes a different course. He chooses to discuss the question of verifiability of a geometry and the question of meaning by a fictitious dialogue between Reichenbach, who represents the opinion of the physicist Hermann von Helmholtz (1821–1894), and Henri Poincaré (1854–1912) (at a later stage in the dialogue Poincaré is replaced by a non-positivist). This dialogue is related to Reichenbach's account of the historical development of the philosophy of geometry: "But the man to whom we owe the philosophical clarification of the problem of geometry is Helmholtz. He saw that physical geometry is dependent on the definition of congruence by means of the solid body and thus arrived at a clear statement of the nature of physical geometry, superior in logical insight to Poincaré's conventionalism, developed several decades later."[33] In contrast to Kant's understanding of space as an a priori notion and to Helmholtz's view that the geometry of space can be empirically determined, Poincaré held that there are some aspects of a scientific theory that can be fixed by convention so that selecting a geometry becomes a matter of choice.

At the beginning of the dialogue, "Poincaré" claims that there are no rigid bodies that can be used for measuring geometric intervals, and therefore geometrical theorems are not verifiable. "Reichenbach" admits that there are no rigid bodies, but their flexibility can be accounted for by such physical effects as volume dependence on temperature, elasticity, and electrostriction. To this "Poincaré" responds that this suggestion to get a real definition of distance is based on physical laws that, in turn, are formulated on the presumption of Euclidean geometry. Hence the verification of geometry is still not possible. He therefore asks why should he not be free to choose a geometry according to his convenience and base the laws of physics on this choice. "Reichenbach" finds this argument attractive, but on the other hand, he suggests that we should continue, as in prerelativistic physics, at least tentatively with the concept of the measurable length as if measuring rigid rods were available. He points out that Einstein could formulate the general theory of relativity only because he adhered to the objective meaning of length. Poincaré would have chosen conventionally Euclidean geometry on account of simplicity. But what matters is the simplicity of all of physics and not of geometry alone, and for that reason one should not adhere to Euclidean geometry. At this stage in the dialogue, out of respect for Poincaré, Einstein replaces him with an anonymous non-positivist. The "non-positivist" questions the notion of distance as a legitimate concept. He invokes the principle (expounded in Reichenbach's essay) that a concept acquires a "meaning" if it is verifiable and applies it to geometrical concepts. He asks how these concepts can have a meaning, if they acquire a meaning only within a fully developed theory however they exist before that stage of the theory.

In this dialogue, Einstein attributes strong arguments to both sides. He settles with Reichenbach on the need for introducing geometrical concepts in their own right but sides with the non-positivist on a holistic conception of a physical theory such as general relativity. How can these two apparently contradictory positions be reconciled? In the final part of the dialogue, the non-positivist criticizes Reichenbach for not having done justice to the philosophical achievements of Kant. Einstein clarifies once more what he sees as valuable in Kant's legacy: not the assumption that specific categories such as those of Euclidean geometry have an a priori status, but the recognition that, in general, some concepts and categories have to be presupposed in order to set up a physical theory, even if they eventually have to be modified again. He thus insists on a more historical perspective on the conceptual development than either of the two opponents of his fictitious dialogue, but he does not go into any more details.

In the unpublished response, Einstein approvingly commented on Reichenbach's account of the historical development of which the theory of relativity is a kind of culmination point, and he liked the representation of the Kant-Helmholtz contradicting views on the cognition of space.[34] In the published text, Einstein concludes his remarks on Reichenbach's essay with a hint to the open nature of the problems discussed, expressing his appreciation: "I can hardly think of anything more stimulating as the basis for discussion in an epistemological seminar than this brief essay by Reichenbach (best taken together with Robertson's essay)."[35]

C. RESPONSE TO PERCY BRIDGMAN

Reichenbach's article followed by Einstein's discussion is related to Percy Bridgman's essay, "Einstein's Theories and the Operational Point of View." Einstein, therefore, allows himself to offer only brief remarks. However, the unpublished report, which we summarize here, is more detailed.[36] It begins with an assessment of the "operational point of view." Operationalism, as advocated by Bridgman, is based on the notion that a concept has a meaning only if accompanied by a prescribed method of measurement. According to Bridgman, a concept is nothing more than a set of operations. This point of view seems to Einstein both fruitful as well as intolerable. Fruitful insofar as it forces one to a critical attitude toward fundamental concepts and definitions that are being used in the theory, but not tolerable because it ignores the fictitious character of any conceptualization.

Bridgman begins his essay by stating that Einstein did not carry over to the general theory of relativity the lessons and insight of his special theory. What he means is that the basic concepts encountered in the special theory of relativity are operationally well defined, but not the concepts of the general theory. Einstein points out that Bridgman's favorable reference to the definition of simultaneity is based on a simplified definition, which Einstein later recognized as being oversimplified.

To correct this point, Einstein explains that the term of simultaneity of distant events depends on the terms "rigid bodies," "inertial system," and "temporarily sharp light signal." These terms are, however, of a fictional character because it seems to be impossible to create a comprehensive, "operational" basis for the two last terms. Nevertheless, the definition of simultaneity is convincing if one assumes that these last terms are less problematic than the one of simultaneity of distant events. "Insofar, but only insofar, the result . . . of this reasoning seems operationally justified." Einstein concludes that it is wrong to claim that a single theoretical concept or statement is legitimate only if its accuracy can be measured through experimental manipulation.

What seems to Einstein a justified requirement for operationalism is the demand that the theory *as a whole* provides controlled and clear statements about experimental facts. But this request should not be required of each term and statement that occur in the theory. The application of the principle of operationalism to a theory is insofar fruitful, as it intends to explore exactly the relations of the theoretical construct with experience. To his mind, there is actually no physical theory that strictly meets the requirements of the principle of operationalism. General relativity is no exception in that sense.

This discussion is summarized briefly in the published "Reply." In order to be able to consider a logical construct as a physical theory, it is not necessary to demand that all its single assessments can be operationally tested. This has not been achieved by any theory. What is required is that a physical theory provides general assertions that can be tested empirically.

D. RESPONSE TO HENRY MARGENAU

Einstein included Margenau in the group of his colleagues, physicists to whom he addressed the remarks about his conviction that the wave function of quantum mechanics is not a complete description of a physical system. Nevertheless, he also wrote a specific, relatively long, response to Margenau's critical essay—first, a collection of unpublished remarks and then a somewhat different text, which he included in the published version of his "Reply to Criticisms." The concept of "reality," or the "real," is a central theme in the philosophy of science and was already discussed in the context of quantum mechanics and in Einstein's response to Reichenbach. In Margenau's essay, this concept even appears in the title—"Einstein's Conception of Reality," and Einstein's first remark in the unpublished report refers to Margenau's statement that ". . . the best of modern physics avoids the term (reality) and operates entirely within the realm of epistemology or methodology; leaving it to the spectator to construe the meaning of reality in any way he wishes."[37] On this Einstein comments that physicists never speak of reality, and that it does not matter at all if one speaks about it or not—just as people breathing do not speak about the air they breathe. Physicists operate with concepts, which assist them to find their way through the variety of experiences of the senses. In the case of concepts of a certain kind, one can speak about a "physical representation of reality." It therefore makes little sense to ask whether physics represents reality or not. The only question is of which nature this representation should be.[38]

The published response begins with a comment on Margenau's statement that "Einstein's position cannot be labelled by any one of the current names of philosophic attitudes; it contains features of rationalism and extreme empiricism. . . ."[39] Einstein agrees and explains why the fluctuation between these two extremes is unavoidable in the work of a physicist. The physicist attempts to connect as directly as possible his concepts with the world of experience. In this endeavor he adopts an empirical attitude. He becomes a rationalist when he realizes that there is no logical path from the empirical to the conceptual world. Einstein elaborates on an even broader scope of philosophical attitudes in the work of scientists in his response to Lenzen and Northrop (see below).

Einstein next refers to Margenau's discussion of the concept of objectivity, which begins with the statement, "What makes the theory of relativity extraordinarily important for philosophy is its incisive answer to the problem of objectivity."[40] According to Margenau, Einstein's concept of objectivity relates to the basic form of theoretical statements and not to the sphere of perception. Objectivity becomes equivalent to invariance of physical laws, not to physical phenomena or observations.

Einstein is not convinced by Margenau's treatment of this issue and criticizes the basic assertions in Margenau's presentation. His response is discussed in similar terms in the unpublished and the published replies. "Objectivity" is, according to Einstein, a characteristic property of every physical theory, like Newtonian mechanics, the theory of relativity, and quantum theory. Every assertion of such a theory claims objective meaning, unless one assumes that the same physical situation admits of several forms of description that are equally justified (like the coordinate of a particle). In the latter case, the general laws of the theory have to be valid for every justified description, and then one can attribute to them "objectivity."

Einstein disagrees with the claim that "objectivity" presupposes certain group properties of a theory (a mathematical property of the equations of a theory, related to the invariance of these equations with respect to transformation between different systems of coordinates). For him, the importance of the group properties (or the symmetry) underlying a physical theory lies in the fact that they restrict the scope of admissible laws. We have already discussed this point in Part II, chapter 13.

Einstein also disagrees with the main thesis of the section on "objectivity" in Margenau's essay—namely, that "[t]he laws of physics, which are to remain invariant, are always differential equations."[41] He comments that this sentence should be questioned even when one agrees that space and time form a continuum and that all physical quantities have to be represented by continuous functions of the coordinates. There is no guarantee that invariant, sufficiently restricted statements for a continuum, which do not possess the form of differential equations, do not exist. Einstein elaborates on this point in the unpublished response. He himself has tried to make progress in this direction, but his failure is not evidence that these endeavors are objectively worthless. It is not even clear that there is no alternative to the space-time continuum, although Einstein did not find "another way (but through the differential equations) to express what is without doubt true in General Relativity."

In fact, Einstein took the possibility that a future theory might not be based on the continuum concept quite seriously, as John Stachel has observed.[42] In a letter to H. S. Joachim, Einstein wrote: "In any case, the alternative continuum-discontinuum seems to me to be a real alternative, i.e., there is no compromise here. By discontinuum theory I understand one in which there are no differential quotients. In such a theory space and time cannot occur, but instead only numbers and number fields and rules for the formation of such on the basis of algebraic rules with exclusion of limiting processes. Only success can teach us which way will prevail."[43] As we know from the report by Abraham Fraenkel on a conversation he had with Einstein in 1951,[44] almost at the time Einstein wrote his "Reply to Criticisms," he was then still optimistic that such an algebraic alternative could be found. He nurtured this hope until the end of his life. Fraenkel recalls that he described to Einstein the attitude of the (neo)-intuitionists, which ". . . would mean a kind of atomistic theory of functions, comparable to the atomistic structure of matter and energy. Einstein showed a lively interest in the subject and pointed out that to the physicist such a theory would seem by far preferable to the classical theory of continuity." He was undeterred by the difficulties of such an approach, pointed out by Fraenkel, and "urged mathematicians to try to develop suitable new methods not based on continuity."

The discussion with Einstein motivated Fraenkel to describe the essence of intuitionism in mathematics. He concludes his article on the subject, echoing Einstein's attitude and emphasizing a profound analogy between intuitionistic trends and certain ideas of modern physics: "In particular, theories of continuum, which renounce genuine mathematical continuity in favor of either 'atomistic' attitudes or of the conception of continuum as a medium of 'freely developing' instead of static 'being,' cannot fail to evoke among physicists a longing for a new mathematical analysis fit to deal with such pseudo-continua."[45]

Shortly before his death, in an appendix to the fifth edition of *The Meaning of Relativity* (published posthumously), Einstein writes: "One can give good reason why reality cannot at all be represented by a continuous field. From the quantum phenomena it appears to follow with certainty that a finite system of finite energy can be completely described by a finite set of numbers (quantum numbers). This does not seem to be in accordance with continuum theory, and must lead to an attempt to find a purely algebraic theory for the description of reality. But nobody knows how to obtain a basis of such a theory."[46]

It is appropriate at this point to digress from Einstein's reply to Margenau and refer to his response to Karl Menger's article on the various schemes of modern geometry and to Menger's prediction that "[t]he day may well come when physicists will take advantage of the wide generality and enormous variety provided by the concepts of modern geometry." And more specifically: "The relativity theory of the future may seek to formulate intrinsic relations between the lines of general metric spaces, without reference to any arbitrarily chosen frame."[47] This statement resonates well with Einstein. He responds: "Adhering to the continuum originates with me not in a prejudice, but rises out of the fact that I have been unable to think of anything organic to take its place. How is one to conserve four-dimensionality in essence (or in near approximation) and (at the same time) surrender the continuum?"[48]

Now, back to Margenau. Einstein, furthermore, disagrees with the claim that physical laws expressed by differential equations are "least specific." If that statement could be proven, "the attempt to ground physics upon differential equations would then turn out to be hopeless." In the unpublished reply he concludes this discussion by saying: "The moment you can indeed prove it, I have to give up the theoretical attempts, the way I formulate them at the present, as principally inadequate."

In the published reply, Einstein responds to Margenau's attempt to describe Einstein's notion of reality as that of someone trained in the Kantian tradition. We have already quoted Einstein's response to this (Part II, chapter 3).[49] For the sake of completeness, let us also refer to it here. Einstein asserts that he did not grow up in the Kantian tradition and came to understand what is truly valuable in his doctrine only quite late. It is expressed by the sentence: "The real is not given to us but put to us [*aufgegeben*] as an assignment." Einstein interprets this statement to mean that there is a ". . . conceptual construction, [which] refers precisely to the 'real' (by definition), and every further question concerning the 'nature of the real' appears empty."[50] This statement could have been used by Einstein in his "Reply" whenever the question of reality was brought up.

Einstein's following remarks are concerned with Margenau's discussion of the Einstein-Podolsky-Rosen paradox and the opinion that the quantum mechanical description is incomplete and that its statistical character is based on this incompleteness.[51] Einstein thinks that Margenau's defense of the generally accepted interpretation of the meaning of the ψ function misses the main point. To correct this, he presents his version of Niels Bohr's formulation of the Copenhagen interpretation and then presents his conclusion of the paradox: One has to abandon one of the following two statements: (1) the claim that the ψ function provides a complete description, or (2) that the two widely separated objects are independent of each other. Since Einstein discusses this issue in greater detail in the *Autobiographical Notes* (see Part II, chapter 12), there is no need to repeat this discussion here.

Margenau's essay contains a short section in which he discusses the relation between classical and quantum-mechanical descriptions in order to elucidate the fundamental difference between both. According to Margenau, this difference consists in the different concept of state introduced by quantum mechanics. What he does not discuss in this section, however, is the classical limit of quantum mechanics, that is, why is it that, under certain conditions, physical reality can be more or less accurately described by classical physics. All Margenau says at this point is: "Almost all of useful modern atomic theory belongs to it [i.e., quantum mechanics]; on the classical level it corresponds to ordinary dynamics." This is exactly the entry point for Einstein's discussion: "One more remark to Margenau's Sec. 7. In the characterization of quantum mechanics the brief little sentence will be found: 'on the classical level it corresponds to ordinary dynamics.' This is entirely correct—*cum grano salis*; and it is precisely this *granum salis* which is significant for the question of interpretation."[52]

Then Einstein proceeds to address the problem of the classical limit of quantum mechanics. In 1927, Paul Ehrenfest had dealt with the approximate validity of classical mechanics within quantum mechanics. He had essentially shown that Newton's second law holds in the sense of averages taken over the wave packet. Einstein now argues that macroscopic bodies behave indeed for a while as would be expected by classical mechanics but that, after a sufficiently long time, quantum properties prevail so that one can no longer assign a specific location to their center of mass because the wave function of such macroscopic bodies eventually spreads out. But how can this fact be reconciled with our experience that, macroscopically, we can still assign a sharp position to such a body, for instance, by illuminating it against a grid of coordinates? Einstein argues that, from the orthodox quantum mechanical viewpoint, this spreading of the wave function of the macroscopic body must be considered as *real*, while the establishment of its sharp position can only be understood as a property of the combined system of body and illumination. He considers this as another paradox—we would say, at the borderline between quantum and classical physics—just as the paradox of the mark on the paper strip in the earlier example[53] also results from bringing quantum and classical systems into the same picture, a move pioneered by Schrödinger's famous cat thought experiment.

Einstein concludes by admitting that these considerations may appear to be splitting hairs but insists that the future of physics may depend on them. To make this point clear, he alludes once more to the Einstein-Podolsky-Rosen paradox by quoting an anonymous "important theoretical physicist" as being inclined to believe in telepathy, clearly a belief that Einstein does not share. Einstein's discussion with colleagues about problems at the borderline between the classical and the quantum domain continued after the publication of the *Notes*, in particular also with Born and Pauli. Einstein contributed another such example to the Festschrift for Max Born, discussing a macroscopic ball bouncing between two parallel walls from both a classical and a quantum point of view. Until the end of his life, Einstein did not want to give up an empirical realism that he once described to Abraham Pais in the sense that "the moon is also there when nobody looks."[54]

E. RESPONSE TO VICTOR LENZEN AND FILMER NORTHROP

Einstein appreciates the successful efforts of Lenzen and Northrop to treat systematically his "occasional utterances" of epistemological contents. Lenzen turns them into a synoptic total system, and Northrop produces from them a masterpiece of a comparative critique of the major empirical systems. Initially, Einstein wrote a brief remark about Lenzen's article, which he sent to Schilpp, who wrote back to him not wanting to influence Einstein's opinion but asking him to reconsider his response.[55] Einstein was astonished, since he believed that his short reply was quite positive. Nevertheless, he amended his short statement with a more general addition. In the published "Reply," he addressed this expanded statement jointly to Lenzen and Northrop. It is here that he expressed his view on the relation between science and epistemology, which we quote as the epigraph to chapter 3 of Part II: "Epistemology without contact with science becomes an empty scheme. Science without epistemology is—insofar as it is thinkable at all—primitive and muddled." Yet Einstein draws a distinction between the epistemologist and the scientist. The latter cannot afford to carry his quest for an epistemological system too far: ". . . the external conditions, which are set for him by the facts of experience do not permit him to let himself be too much restricted in the construction of his conceptual world by the adherence to an epistemological system." Einstein must refer to himself when he enumerates the different epistemological categories that the scientist may represent: he is a *realist* when he seeks to describe a world, independent from the acts of comprehension; he is insofar an *idealist*, when he regards the concepts and theories as free inventions of the human brain; he is also a *positivist* when he considers his concepts and theories as valid, because they reveal a logical presentation of relations between the sensual experiences. He may even appear as a *Platonist* or *Pythagorean* when he believes in logical simplicity as a guideline in his research.

F. RESPONSE TO ARTICLES ON GENERAL RELATIVITY AND COSMOLOGY (EDWARD MILNE, LEOPOLD INFELD, AND GEORGES LEMAÎTRE)

Edward Milne's article, "Gravitation without General Relativity," presents a theory that does not regard all observers, that is, all reference frames, as equivalent. Einstein's unpublished commentary begins with remarks on the first sentence in this article: "General relativity arose through the supposed impossibility of bringing gravitation within the scope of the Lorentz formulae of what has been called 'Special Relativity.'" Einstein disagrees with this statement. As he points out in the unpublished reply,[56] the emergence of the general theory of relativity is based on three arguments:

1. The equality of gravitational and inertial mass suggests a theory that in its very foundation does justice to this elementary fact. This cannot be achieved within the framework of special relativity theory in a satisfactory manner.
2. If one conceives of gravitation systematically as a field, then there is no possibility to empirically recognize an inertial system as such.
3. Privileging inertial systems over other, from a kinematic point of view, equivalent systems is per se unsatisfactory, as Newton and Mach already clearly recognized.

Finally, he remarks that it is unnatural to restrict the legitimate transformations of coordinates to Lorentz transformations (the special theory of relativity) when one abandons the viewpoint that the space and time coordinates are directly measurable by rods and clocks.

Einstein concludes by asserting that one cannot achieve a reliable theory of the structure of the physical continuum without grasping the deeper reason for the equality of the inertial and gravitational masses. In the published reply he makes a similar remark. He believes that ". . . one cannot arrive, by way of theory, at any at least somewhat reliable results in the field of cosmology, if one makes no use of the principle of general relativity."

On Leopold Infeld's essay, "General Relativity and the Structure of Our Universe," Einstein comments briefly that it is an "independently understandable, excellent introduction to the so-called 'cosmological problem' of the theory of relativity, which critically examines all essential points." In the unpublished response he adds a few remarks.[57] He agrees that the simple solution with zero curvature is not satisfactory as a hypothetical description of the universe. What Infeld's article demonstrates clearly is that a cosmological model with zero curvature alone (without a cosmological constant) is enough to describe a universe with finite density of matter, but that is not the case for non-zero curvature without expansion. However, a careful discussion of the empirical determination of the expansion (Hubble) constant is desired, because there is a discrepancy between the value obtained from the cosmological model implied by the theory of general relativity and from other empirical data. Einstein returns to this point in his reply to Lemaître.

Georges Lemaître in his article on the cosmological constant insists ". . . on the logical convenience or even the theoretical necessity of its introduction." Einstein introduced the cosmological constant into the gravitational field equation in his first paper on the cosmological consequences of the general theory of relativity in 1917. This was necessary to formulate his model of a static universe with an average homogenous density of matter. Einstein never liked this modification of the gravitational field equation because it "constitutes a complication of the theory, which seriously reduces its logical simplicity."[58] After Hubble's discovery of the expansion of the universe, Einstein dropped the cosmological constant and adopted the model of Alexander Friedmann, who showed in 1922 that models of an expanding universe without a cosmological constant and with finite density of matter exist. Einstein described this development in the appendix to the second edition, 1945, of his *Meaning of Relativity*.[59]

There remained, however, a problem with the estimated age of the universe derived from Hubble's rate of expansion (Hubble's constant). A reliable estimate of the age of the earth's crust exceeded the estimated age of the universe based on the expansion rate, of about 10^9 years. This was also incompatible with the contemporary astrophysical theories of the evolution of stars, which require longer times.

Einstein nevertheless responds to Lemaître that the introduction of the cosmological constant was not justified at that stage of knowledge. Moreover, referring to the problem of the estimated age of the universe, Einstein claims that it would not offer an "absolutely natural escape from this difficulty." He was prepared to blame the problem of the age of the universe on the fact that general relativity is an incomplete theory (i.e., not yet a unified field theory), rather than on the absence of the cosmological constant in his original field equations.

In his reply to Lemaître he reviews the basic tenets of general relativity:

1. Physical concepts are described by continuous functions of four coordinates; they can be freely chosen, as long as their continuity is preserved.
2. The field variables are components of tensors; there is a symmetric tensor for the description of the gravitational field.
3. There are physical objects that measure the invariant distance ds between two adjacent points in space-time.

Einstein then points out that the use of physical objects that measure the invariant distance between two points is a "psychologically" important auxiliary construction, which a final theory should eliminate by incorporating such objects within the theory. He then vaguely hints at the possibility that such a modified theory might give rise to time measurements that would remove the discrepancy in the estimated age of the universe.

The nature of rods and clocks was first discussed by Hermann Weyl in his proposed theory unifying the gravitational and electromagnetic fields. We have discussed this theory and its implications for space and time measurements in our *Formative Years of Relativity*.[60] Weyl's theory implied that the measuring devices—rods and clocks—lose their invariance and depend on the history of their movement in space-time. If this were the case, then two close identical atoms would emit light at slightly different frequencies, and we would not observe sharp spectral lines. Atoms could no longer serve as standard clocks, and the frequency of light emitted from atoms in distant galaxies could not be used to measure time intervals. Weyl argued that the assumption that intervals in space-time can be measured directly with rods and clocks is problematic, and that rods and clocks should be dispensed with as measuring devices. At the time, Einstein objected to Weyl's theory on the grounds of this possible consequence, but now he clings to this straw, in order to justify his objection against introducing the cosmological constant.

Einstein's response to Lemaître is astonishing. He evidently prefers this quite speculative possibility over the straightforward solution that the introduction of a cosmological constant would have offered him. Actually, the question was not of introducing or removing the cosmological constant. It is a term in the gravitational field equation, and the question was merely what value should be assigned to it. In 1931, the physicist and cosmologist Richard Tolman suggested to Einstein that there is a strong argument against assigning the value zero to the cosmological constant.[61] More than fifteen years later, Einstein still stubbornly objected to it.

The problem of the age of the universe is related to measurements of Hubble's constant. Its value was corrected, only a few years after the publication of Schilpp's volume, by work on stellar populations.

G. RESPONSE TO KURT GÖDEL

In the late 1940s, the mathematician Kurt Gödel found an unusual solution to the field equations of general relativity, describing a rotating universe. Gödel showed that two types of such solutions exist—solutions describing a static and an expanding universe. The latter type is more relevant to our universe. A unique feature of these solutions is that they contain closed world lines extending over large domains of space-time. On such a world line, it is not possible to assign uniquely a past-future relation between two distant points. In a short essay (the shortest in the volume), titled "A Remark about the Relationship between Relativity Theory and Idealistic Philosophy," Gödel uses this result to discuss the, apparently, paradoxical possibility of time travel to the past and, more important, its philosophical implications in support of an idealistic, specifically Kantian, philosophy in which time is not a physical entity but a basic a priori concept in our thinking process. Gödel had already reached this conclusion in the context of special relativity on the basis of the relativity of simultaneity and on the fact that different observers will see different successions of "nows" between two points in space-time. For Gödel, this means that there is no objective lapse of time, favoring the idealistic perception that time does not have a real physical meaning.

In his responses to other contributors in this volume, Einstein does not hesitate to criticize arguments that, according to him, are based on a misunderstanding. Maybe out of respect for Gödel, with whom he spent many hours walking and talking in Princeton, Einstein tactfully ignores the contrived philosophical arguments in Gödel's essay, which may either reflect a misunderstanding of special relativity or constitute a metaphysical addition.

Einstein refers only to Gödel's mathematical results, which he considers to be an important contribution to the analysis of the concept of time in the general theory of relativity. He wishes to explain to the reader, in simple terms, the meaning of Gödel's unique solutions of the gravitational field equations. Einstein begins with a discussion of the direction of time in the neighborhood of a point P ("event") in space-time. To every point belongs a region of space-time enclosed by a cone (the "light cone") such that every point in that region is connected to point P by a possibility of exchanging light signals between P and that point. Looking at two points, A and B, on both sides of P, connected by a "timelike" line (a line along which it is possible to send a signal), there is no free choice of the direction of flow of time. Sending a light signal is an irreversible process; the points are connected by light signal causality, and therefore the statement "point B precedes point A" has a physically unambiguous sense.

The situation is more complicated when the two points are separated by a large, say cosmological, distance from each other and still connected by a timelike line. That line is a series of infinitesimal time lapses with a definite direction of the time arrow. However, what happens if this series of time lapses is closed on itself? When the two points are close to each other, the statement that "B precedes A" still makes physical sense. But if they are cosmologically far apart it is meaningless to say which of the two points precedes the other, giving rise to all the paradoxes related to the loss of causal connection between them.

At the end of the handwritten version of his "Reply to Criticisms," Einstein devotes a paragraph, which he erased, to the discussion of the "arrow nature" of time, his term for Gödel's "flow of time."[62] The "arrow nature" of time is related to the second law of thermodynamics. It could be extended over arbitrarily large space-time domains if the gravitational field equations would not allow closed timelike world lines. This is what Einstein assumed until the cosmological solutions found by Gödel showed that this assumption

Einstein and Gödel—
disappearance of the
arrow-nature of time.

was wrong. In the erased paragraph, Einstein states that he sees no justification to exclude a priori such solutions despite their paradoxical implications caused by the loss of the "arrow nature" of time. In the published version of his response, Einstein says that Gödel's solutions are important to the general theory of relativity and concludes that "It will be interesting to weigh whether these are not to be excluded on physical grounds." He does not suggest what such physical grounds could be.

NOTES

1. Einstein to Schilpp, 29 May 1946, AEA 42-514.
2. See Einstein's "Reply to Criticism," in Schillp, ed., *Albert Einstein: Philosopher-Scientist*, p. 665.
3. Einstein, unpublished response, AEA 2-060.
4. Ushenko to Einstein, 5 September 1949, AEA 2-061.
5. Einstein to Ushenko, 6 September 1949, AEA 2-062.
6. Schilpp to Einstein, 28 March 1949, AEA 80-524.
7. Einstein, unpublished response, AEA 2-064.
8. Einstein, unpublished response, AEA 2-035.
9. Einstein, unpublished response, AEA 2-049.
10. Einstein to Schilpp, 20 June 1949, AEA 80-490.
11. Schilpp to Einstein, 26 June 1949, AEA 80-529.
12. Einstein to Schilpp, 28 June 1949, AEA 80-491.
13. Einstein to Schilpp, 25 March 1949, AEA 80-485.
14. Einstein to Johanna Fantova, 19 February 1949, AEA 87-318.
15. Einstein, unpublished response, to Born AEA 2-028 and to Heitler AEA 2-037.
16. Albert Einstein, "Ernst Mach," *Physikalische Zeitschrift*, 17 (1916): 101–104, here 101. In CPAE vol. 6, Doc. 29.
17. Born in Schilpp, ed., *Albert Einstein: Philosopher-Scientist*, p. 176.
18. Max Born, *The Born-Einstein Letters: Friendship, Politics, and Physics in Uncertain Times: Correspondence between Albert Einstein and Max and Hedwig Born from 1916 to 1955 with Commentaries by Max Born* (New York : Macmillan, 2005), 145.
19. Einstein, unpublished response, AEA 2-028.
20. Born in Schillp, ed., *Albert Einstein: Philosopher-Scientist*, p. 176.
21. Einstein, unpublished response, AEA 2-037.
22. Einstein's "Reply to Criticism," p. 667.
23. For Einstein's views on quantum mechanics, see also Christoph Lehner, "Realism and Einstein's Critique of Quantum Mechanics," in *The Cambridge Companion to Einstein*, ed. Michel Janssen and Christoph Lehner (Cambridge: Cambridge University Press, 2014).
24. Einstein's "Reply to Criticism," p. 671.
25. Ibid., p. 672–673.
26. Ibid., p. 674.
27. Einstein in Paul Arthur Schilpp, ed., *The Philosophy of Bertrand Russell*, 291.
28. Einstein's "Reply to Criticism," p. 674.
29. Ibid., p. 676.
30. Einstein, unpublished response, AEA 2-058.
31. Reichenbach in Schillp, ed., *Albert Einstein: Philosopher-Scientist*, p. 293.
32. Ibid., p. 294.
33. Ibid., p. 300.
34. Einstein's attitude to Reichenbach's philosophy, including this dialogue, is discussed by Don Howard in "Einstein and the Development of Twentieth-Century Philosophy of Science," in *The Cambridge Companion to Einstein*, ed. Michel Janssen and Christoph Lehner (Cambridge: Cambridge University Press, 2014).
35. Einstein's "Reply to Criticism," p. 679.
36. Einstein, unpublished response, AEA 2-030.

37. Henry Margenau, "Einstein's Conception of Reality," in Schillp, ed., *Albert Einstein: Philosopher-Scientist*, p. 248.

38. Einstein, unpublished response, AEA 2-042.

39. Margenau in Schillp, ed., *Albert Einstein: Philosopher-Scientist*, p. 247.

40. Ibid., p. 252.

41. Ibid., p. 254.

42. John Stachel, "The Other Einstein: Einstein contra Field Theory," *Science in Context* 6, no. 1 (1993): 275–290.

43. Einstein to H. S. Joachim, 14 August 1954, AEA 13-454.

44. Stachel, "The Other Einstein," 287–288.

45. Abraham H. Fraenkel, "The Intuitionistic Revolution in Mathematics and Logic," *Bulletin of the Research Council of Israel* 3 (1954): 283–289.

46. Albert Einstein, *The Meaning of Relativity*, 5th ed. (Princeton, NJ: Princeton University Press, 1956).

47. Karl Menger, "Theory of Relativity and Geometry," in Schillp, ed., *Albert Einstein: Philosopher-Scientist*, 471.

48. Einstein's "Reply to Criticisms," p. 686.

49. See also Thomas Ryckmann, "'A Believing Rationalist': Einstein and 'the Truly Valuable' in Kant," in *The Cambridge Companion to Einstein*, ed. Michel Janssen and Christoph Lehner, 377–397 (Cambridge: Cambridge University Press, 2014).

50. Einstein's "Reply to Criticisms," p. 680.

51. Einstein, Podolsky, and Rosen, "Can Quantum-Mechanical Description of Physical Reality Be Considered Complete?"

52. Einstein's "Reply to Criticisms," p. 682.

53. Ibid., p. 670.

54. Quoted in Christoph Lehner, "Realism and Einstein's Critique of Quantum Mechanics," in *The Cambridge Companion to Einstein*, ed. Michel Janssen and Christoph Lehner (Cambridge: Cambridge University Press, 2014), 344.

55. Schilpp to Einstein, 29 April 1947, AEA 42-518.

56. Einstein, unpublished response, AEA 2-046.

57. Einstein, unpublished response, AEA 2-040.

58. Einstein, appendix to the 2nd ed. of *The Meaning of Relativity*, in Hanoch Gutfreund and Jürgen Renn, *The Formative Years of Relativity: The History and Meaning of Einstein's Princeton Lectures* (Princeton, NJ: Princeton University Press, 2017), 271.

59. See Gutfreund and Renn, *The Formative Years of Relativity*, Part II, chap. 5.

60. Ibid., chap. 8.

61. Tolman to Einstein, 14 September 1931, AEA 23-031.

62. Einstein, handwritten version of the Reply to Criticisms, AEA 2-025.

PART IV

EINSTEIN'S "AUTOBIOGRAPHICAL SKETCH" (1955)

1
INTRODUCTORY REMARKS

Einstein was approached in August 1954 with a request to contribute a few pages of his reminiscences to be published in a festive issue of *Schweizerische Hochschulzeitung* (Swiss Universities Journal) on the occasion of the hundredth anniversary of the Federal Technical University (ETH) in Zurich to take place in October 1955. Einstein responded only at the end of February 1955 after receiving a reminder of the initial request. He explained his long silence by the time it took him to decide how to best satisfy the friendly offer extended to him. He asked if an account of his collaboration with his friend from student days and later professor at the ETH, Marcel Grossmann, would be appropriate. Einstein sent his completed manuscript on 29 March, about two weeks before his death. As the last essay written by Einstein, it is of great interest. It was published in the jubilee book of the ETH and also included in a memorial volume edited by Carl Seelig, under the title "Autobiographical Sketch."[1]

Einstein explained his agreement to contribute to the jubilee book: "The urge to express my gratitude to Marcel Grossmann at least once in my lifetime is what gave me the courage to write this somewhat colorful autobiographical sketch." Einstein and Grossmann became close friends as students at ETH. Grossmann studied mathematics and in 1907 became professor of descriptive geometry at the same university. In the following years, he not only helped Einstein in his academic career but also in his scientific work. Einstein truly considered Grossmann as his friend. In 1905, he dedicated his doctoral dissertation, "A New Determination of Molecular Dimension," to "My Friend Dr. Marcel Grossmann."

In 1911, Marcel Grossmann was appointed dean of the mathematics-physics department of ETH. One of his first initiatives was to write to Einstein asking if he would be interested in returning to Zurich to join the ETH. Einstein agreed, although he could have expected a similar offer from Holland, where he would have eventually been designated to succeed Lorentz. Whatever the reasons for preferring Zurich may have been, at that time it was the right decision. Einstein was then serving as professor of theoretical physics at the German part of the Charles University in Prague. He was working on the general theory of relativity and soon realized that more sophisticated mathematical methods than he was familiar with at that time would be required to make progress. He then turned to his friend the mathematician: "Grossman, you must help me or I'll go crazy."[2]

A short time after arriving in Zurich in August 1912, Einstein began an intensive and fruitful collaboration with Grossmann, which became a landmark in the development of general relativity. Grossman introduced Einstein to recent developments in the absolute differential calculus by Riemann, Ricci-Curbastro, and Levi-Civita. Their collaboration

Classmates (*left to right*): Marcel Grossmann, Albert Einstein, Gustav Geissler, and Eugen Grossmann in the garden of the Grossmann family house near Zurich/Thalwil, May 1899. © Hebrew University of Jerusalem, Albert Einstein Archives, courtesy AIP Emilio Segre Visual Archives.

is best reflected in the "Zurich Notebook," which eventually led to the joint publication of the so-called *Entwurf* theory (see box). The problem at the core of Einstein's research documented in this notebook was to find a field equation for the gravitational field, namely, to find a relation that determined how this field is generated by its source, that is, by energy and matter. The notebook contains an important entry connected with the assistance of Marcel Grossmann. Grossmann referred Einstein to a key mathematical concept, the so-called Riemann tensor, and thereby showed him the royal road to general relativity from the point of view of the present-day theory. The "Zurich Notebook" contains the essentially correct field equation in a weak field approximation, but owing to unresolved conceptual difficulties it was abandoned. It was revisited in November 1915, when the final version of the general theory of relativity was completed.

THE *ENTWURF* THEORY

In the spring of 1913, Einstein derived, with Grossmann's help, a gravitational field equation that became known as the core of the *Entwurf* (draft) theory. It primarily satisfied requirements rooted in classical physics, namely, the energy-momentum conservation, and the expectation that it reduces to Newton's theory in the limit of weak static gravitational fields. To satisfy these requirements, as he then understood them, he had to impose certain restrictions on admissible coordinates. As a result, the class of coordinate systems in which the *Entwurf* equation takes on the same form does not satisfy the generalized principle of relativity in the way he imagined. He therefore abandoned with a heavy heart the realization of general covariance. From the modern perspective, this theory is incorrect, but at that point in time Einstein assumed that it was the best that could be done, and he even convinced himself that a gravitational field theory does not allow general covariance. In late 1915, Einstein realized the fallacy of this conclusion. This insight led him, in a short period of time, to his final formulation of the general theory of relativity.

Although the *Entwurf* theory is not correct, it has been considered by historians of science as the scaffold on which the general theory of relativity could be constructed.[3]

Looking back at the years of friendship and scientific collaboration with Grossmann, toward the end of his life, Einstein thought that he did not thank him properly in the past, and he wanted to make up for it on this occasion. Einstein was, in general, not generous with referring to other people's work and acknowledging their contributions, but ignoring Grossmann's role in several paradigmatic publications must have generated, at this stage of his life, a sense of regret.

Einstein did not mention Grossmann in his November 1915 papers, submitted to the Royal Prussian Academy of Sciences, in which he presented the final formulation of the theory. He did acknowledge Grossmann's contribution in the review paper of 1916, "The Foundation of General Relativity," but apparently only as a second thought. Looking at the manuscript, we can see that he first wrote the titles of the first chapter and section, then shifted them to the bottom of the page in order to include some introductory remarks, among them: "Finally, I want to acknowledge gratefully my friend, the mathematician Marcel Grossmann, whose help not only saved me the effort of studying the pertinent mathematical literature, but who also helped me in my search for the field equations of gravitation."

Einstein did not mention Grossmann in his *Relativity—The Special and General Theory (a Popular Account)*. This book had been translated in the 1920s into ten languages. To a few of the foreign editions Einstein added a brief introduction. He wrote, for example, an introduction to the Czech translation. He did it with pleasure, remembering the days he spent in Prague. In the introduction he recalled: ". . . the decisive idea of the analogy between the mathematical formulation of the theory and the Gaussian theory of

First page of the manuscript "The Foundation of General Relativity" with an acknowledgment of Marcel Grossmann. © Hebrew University of Jerusalem.

surfaces came to me only in 1912 after my return to Zurich, without being aware at that time of the work of Riemann, Ricci and Levi-Civita. This was first brought to my attention by my friend Grossmann. . . ." Einstein did not mention Grossmann in his canonical book *The Meaning of Relativity*, nor in his *Autobiographical Notes*. In the "Autobiographical Sketch," he compensated for these omissions with grace. We do not know if Grossmann developed bad feelings against his friend. It was, however, too late for him to enjoy Einstein's appreciative remarks. After their collaboration on the general theory of relativity between 1912 and 1914, they had again a scientific exchange in the early 1930s, this time concerning Einstein's approach to a unified field theory on the basis of teleparallelism. In 1931, Grossmann even published an article critical of Einstein in which he wrote: "Also earlier Einstein has published—it was in the year 1913—field equations according to this method, which had to be modified after a few years; at the time I shared the responsibility." This time he felt the urge to warn against this development. This exchange did, however, not hamper their friendship. When Grossmann died in 1936 after many years of suffering from multiple sclerosis, Einstein wrote to his widow: "One thing however is beautiful: We were and remained friends throughout life."[4]

In the texts summarizing Einstein's road to general relativity, mentioned above, he never referred to the intermediate step on this road, which has been known as the *Entwurf* theory of 1913 (see box). It would have been natural for him to do so in the "Autobiographical Sketch," where he described his work with Grossmann in greater detail than ever before. But even there, he did not mention this publication. He only remarked that the erroneous considerations in 1912 deferred the completion of the theory until 1916 (should be 1915).

In the last paragraph of this essay, Einstein remarks on his contemporary thinking about the forty-year effort in search of a field theory, based on a generalization of his theory of gravitation, that could form a foundation of the entire physical world. He concludes this discussion expressing the same doubt and using almost the same words that he used at the end of the *Autobiographical Notes*, about nine years earlier: "It is doubtful if a field theory can account for the structure of matter, of radiation and of quantum phenomena." If "field" is understood here as a classical continuum, the majority of physicists would answer this query with a convinced "NO," believing that quantum phenomena have to be accounted for by other methods. In the past, Einstein used to counter this with his conviction that it was possible and would ultimately be achieved. At this stage he only quotes Lessing's consoling words: "The quest for truth is more delightful than its assured possession."

Last known photograph of Albert Einstein, taken on his birthday, March 14, 1955. Photo by Ardon Bar Hama. Courtesy of the Albert Einstein Archives.

NOTES

1. Albert Einstein, "Erinnerungen-Souvenirs," *Schweizerische Hochschulzeitung* 28 (Sonderheft) (1955): 145–148, 151–153. Reprinted as "Autobiographische Skizze," in Carl Seelig, *Helle Zeit—Dunkle Zeit: In Memoriam Albert Einstein* (Zurich: Europa, 1956), pp. 9–17.

2. "Grossmann, Du mußt mir helfen, sonst werd' ich verrückt!" Cited in Louis Kollross, "Erinnerungen eines Kommilitonen," in *Helle Zeiten—Dunkle Zeiten: In Memoriam Albert Einstein*, ed. Carl Seelig (Zurich: Europa Verlag, 1955), 17–31, here 27.

3. See Albert Einstein and Marcel Grossmann, "Outline of a Generalized Theory of Relativity and of a Theory of Gravitation" (1913), in *The Collected Papers of Albert Einstein*, vol. 4, Doc. 13, pp. 151–188; Michel Janssen and Jürgen Renn, "Arch and Scaffold: How Einstein Found His Field Equations," *Physics Today* 68 (November 2015): 30–36.

4. Quoted in Claudia E. Graf-Grossmann, *Marcel Grossmann: For the Love of Mathematics* (Cham, Switzerland: Springer, 2018), 166.

2

"AUTOBIOGRAPHICAL SKETCH"

AN ENGLISH TRANSLATION

IN 1895, AGED SIXTEEN, I ARRIVED IN ZURICH FROM ITALY AFTER SPENDING A YEAR AT my parents' home in Milan without attending school or having any tuition at all. My goal was admission to the Polytechnic, but I had no clear idea for achieving it. I was a headstrong yet modest young person who had acquired the pertinent elements of his patchy knowledge mostly by studying on his own. While I was keen to delve deeper, I lacked the talent for absorbing knowledge and was hampered by a bad memory, so university studies seemed far from easy to me. I was right to feel hesitant as I registered for the entrance examination in the engineering department. Although the examination made me painfully aware of the gaps in my education, the examiners were patient and understanding. My failure seemed completely justified, but it was comforting that the physicist H. F. Weber let me know I could attend his lectures if I stayed in Zurich. Meanwhile the rector, Professor Albin Herzog, recommended me to the cantonal school in Aarau, where I studied for a year and received my high school graduation certificate. This school, with its liberal spirit and the unpretentious, serious attitude of the teachers, who relied on their own judgment rather than any outside authority, made a lasting impression on me. Comparing this with my six years of education at a German high school run by authoritarian methods made me acutely aware that it is far better to teach people to act freely and on their own responsibility than to educate them on principles of military drill, external authority, and ambition. Genuine democracy is not an empty illusion.

During the year in Aarau, the following question occurred to me: If one chases a light wave at the speed of light one would arrive at a time-independent wave field. But nothing like that really seems to exist! This was the first childlike experiment in thinking about special relativity theory. Innovation itself is not the result of logical thought, even though the end product is tied to a logical structure.

From 1896 to 1900 I studied at the specialist teachers' department at the Federal Polytechnic. Before long I realized that I had to be content with being an average student. To

Translated from the German by Karen Margolis.

be a good student you must be able to grasp things easily; you must be willing to concentrate your energies on everything you are told in the lectures; you must enjoy writing down everything presented in the lectures in an orderly fashion and working on it conscientiously. Regretfully, I realized that I fundamentally lacked all these qualities. This meant that I gradually learned to live in peace with a degree of bad conscience and to organize my studies to suit my intellectual tastes and my own interests. I followed some of the lectures with interest and excitement. Otherwise, I often skipped classes and studied the masters of theoretical physics at home with a divine zeal. This was a good thing in itself and soothed my bad conscience so effectively that I avoided any serious emotional upsets. I resumed my earlier habit of long private study sessions, in which I was joined by a Serbian student, Mileva Marić, whom I later married. At the same time, I worked zealously and passionately in Professor H. F. Weber's physics laboratory. I was also fascinated by Professor Geiser's lectures on infinitesimal geometry, which were veritable masterpieces of the art of education and proved very helpful later when I was wrestling with the general theory of relativity. Aside from this, higher mathematics was of little interest to me during my university studies. I mistakenly believed that this was such a widespread and diffuse territory that one could easily waste all one's energy in a remote province. In my innocence I thought it was enough for a physicist to have grasped the elementary mathematical concepts clearly and to be able to use them readily, and that everything else involved subtleties that were unproductive for physicists—an error I regretfully realized only later. Evidently I lacked sufficient mathematical talent to be able to distinguish between central, fundamental matters and peripheral things of no major importance.

In my university days I became close friends with a fellow student, Marcel Grossmann. I had a regular weekly date with him in Café Metropol on Limmatquai, and we talked not only about our studies but also about everything else of possible interest to curious, open-minded young people. Unlike me, he was not a vagabond and a loner type; he was someone who was deeply rooted in his Swiss surroundings yet never lost his innate independence. He was also richly endowed with all the gifts I lacked: he was quick to grasp things and orderly in every sense. He attended all the lectures that interested us, and his work on them was so excellent that his notebooks were eminently fit for publication. He lent me those notebooks to prepare for the examinations, and they were like a life belt for me. I do not care to speculate how things would have turned out for me without them.

Even with this invaluable help, and although the subjects discussed in our lectures were intrinsically interesting, I still had to struggle to overcome my reluctance to learn all these things thoroughly. Academic studies are not necessarily beneficial for withdrawn, contemplative people like myself. Constantly being forced to eat so many good things can ruin your appetite and give you bellyache. It can extinguish the delicate light of divine curiosity forever. Fortunately, in my case my studies ended well, and this intellectual depression lasted only a year after I finished university.

The best thing Marcel Grossmann ever did for me as a friend happened around a year after I finished studying. With his father's help he recommended me to the director of the Swiss Patent Office, which was still called the Office of Intellectual Property in those days. After an extensive interview Mr. Haller gave me a job there. This relieved me of existential pressures during the years when I produced my best work. Aside from this, working to evaluate technical patent applications was a real blessing for me. It compelled me to think in many different ways and provided important stimuli for my thinking about physics.

A practical profession is ultimately a blessing for people like myself. An academic career puts a young person into a coercive situation in which he is compelled to produce impressive quantities of scientific papers, and this creates a temptation to superficiality that only strong-minded characters can resist. What is more, in most practical professions an averagely gifted person can perform according to expectations. His life in society is not dependent on any special intellectual discoveries. If his scientific interests go deeper, he can get immersed in his favorite problems alongside his usual work obligations. He has no need to worry that his efforts might fail to yield results. Thanks to Marcel Grossmann, I was now in this fortunate position.

Among the scientific experiences of those happy years in Bern I shall mention one in particular, which turned out to be the most fruitful idea of my life. The theory of special relativity was already a few years old. The question was whether the principle of relativity was limited to inertial systems, that is, coordinate systems that move in a straight line at constant velocity relative to each other (linear coordinate transformations). On a formal level one would instinctively say, "Probably not!" Yet the foundation of every kind of mechanics until then—the principle of inertia—seemed to exclude any extension of the principle of relativity. In fact, if one introduces an accelerated coordinate system (relative to an inertial system), an "isolated" mass point no longer moves uniformly and in a straight line in relation to it. At this juncture, a mind not bound by narrow thinking habits would have asked, "Does this type of motion offer me a means of differentiating between an inertial and a non-inertial system?" The untrammeled mind would have necessarily have answered negatively (at least in the case of uniform acceleration in a straight line). For one could also interpret the mechanical behavior of bodies relative to such an accelerated coordinate system as the effect of a gravitational field; this is possible by virtue of the empirical fact that in a gravitational field, too, the acceleration of bodies is always the same, independent of their nature. This insight (the equivalence principle) not only made it probable that the natural laws must be invariant with respect to a universal group of transformations corresponding to the group of Lorentz transformations (extension of the principle of relativity), but also that this extension would lead to an advanced theory of the gravitational field. I had not the least doubt that this idea was correct in principle. But it seemed nearly impossible to establish this conclusively. To begin with, there were elementary considerations that the transition to a further group of transformations is not compatible with a direct physical interpretation of the space-time coordinates that had prepared the ground for special relativity theory. Moreover, at first it was hard to see how to choose the extended group of transformations. In fact, I arrived at the equivalence principle by way of a detour that is beyond the scope of the present account.

This problem occupied me continuously all through the period 1909–1912, while I was engaged in teaching theoretical physics at the universities of Zurich and Prague. By 1912, when I was appointed as a professor at Zurich Polytechnic, I had come much closer to solving the problem. Hermann Minkowski's analysis of the formal foundations of special relativity theory emerged as an important factor here. It can be summarized in the sentence: Four-dimensional space has an (invariant) pseudo-Euclidean metrics; this determines the experimentally verifiable metrical characteristics of the space as well as the inertia principle and, moreover, the form of Lorentz-invariant equivalence systems. This space contains preferred, namely, quasi-Cartesian coordinate systems, which are the only "natural" ones here (inertial systems).

The equivalence principle leads us to introduce nonlinear coordinate transformations into this type of space, that is, non-Cartesian ("curvilinear") coordinates. In this case, the pseudo-Euclidean metrics assumes the universal form:

$$ds^2 = \Sigma \, g_{ik} \, dx_i \, dx_k$$

summed over the indices i and k (which take values from 1–4). These g_{ik} functions are then functions of the four coordinates that, according to the equivalence principle, describe not only the metrics but also the gravitational field. The latter has a very special quality, of course; as we know, it can be transformed into the special form

$$-dx_1^2 - dx_2^2 - dx_3^2 + dx_4^2$$

that is, a form in which the g_{ik} functions are independent of the coordinates. In this case, it can be "transformed away" by the g_{ik}-defined gravitational field. In the latter special form, the inertial motion of isolated bodies is expressed by a (timelike) straight line. In the universal form, it is expressed by the "geodetic line."

Although this formulation still derived from the case of pseudo-Euclidean space, it clearly showed how the transition to gravitational fields of a general type was to be achieved. Here, too, the gravitational field can be described by a type of metrics, that is, by a symmetrical tensor field, g_{ik}. The generalization merely consists in leaving out the precondition that this field can be transformed into a pseudo-Euclidean one simply by transforming the coordinates.

This solution reduced the problem of gravitation to a purely mathematical one. Are there differential equations for the g_{ik} functions that are invariant with respect to nonlinear coordinate transformations? Such differential equations, and *only* such, would come into consideration as field equations of the gravitational field. The law of motion of material points was then given by the equation of the geodetic line.

It was with this problem in mind that I went to see my old friend from university, Marcel Grossmann, in 1912. By then he had become a professor of mathematics at the Swiss Federal Polytechnic. He caught on to the problem right away, even though as a true mathematician he had a rather skeptical attitude toward physics. In our student days when we used to discuss ideas over coffee, he once made such a lovely characteristic remark that I cannot resist quoting it: "I must admit that studying physics has actually benefited me considerably. Before, when I sat down on a chair that felt a bit warm from the person who had occupied it previously, I used to feel rather uncomfortable. This feeling has completely gone now, because physics has taught me that the heat is something totally impersonal."

In the event, he was fully prepared to collaborate on the problem with me, but only on condition that he would not have to take responsibility for any claims and interpretations related to physics. He checked out the literature and soon discovered that the mathematical problem in question was already solved, in particular by Riemann, Ricci, and Levi-Civita. The whole development connected up with the theory of the Gaussian curvature, which systematically used generalized coordinates for the first time. Riemann's achievement was the greatest. He showed that tensors of second-order covariant differentiation could be formed out of the field of g_{ik} tensors. This revealed what the gravitational field equations must look like—under the condition that invariance is required with respect to the group of all continuous coordinate transformations. It was, however, not very easy to see that this condition was justified, especially as I believed I had found arguments against it. This reservation, which eventually turned out to be mistaken, resulted in the theory only appearing in its final form for the first time in 1916.

While I was avidly working with my old friend, none of us dreamed that this outstanding man would be carried off by a pernicious illness far too early. It was the urge to express my gratitude to Marcel Grossmann at least once in my lifetime that prompted me to write this rather slapdash autobiographical note.

Forty years have passed since the theory of gravitation was completed. Those years were almost entirely devoted to efforts to develop a unified field theory that could form a foundation for the whole of physics by generalizing from gravitational field theory. Many people worked toward the same goal. I investigated several seemingly promising approaches and subsequently discarded them. But the past ten years finally led to a theory that seems natural and promising to me. Still, I have not been able to convince myself about whether I should regard this theory as valuable for physics or not. This is fundamentally due to mathematical difficulties that are insurmountable for the time being, like those, incidentally, that arise in the use of any nonlinear field theory. Moreover, it seems doubtful altogether whether a field theory can properly account for the atomistic structure of matter and radiation as well as of quantum phenomena. Most physicists would immediately answer "no," because they believe that the quantum problem has been solved in principle in another way. Be that as it may, we should take comfort from Lessing's dictum that the search for truth is more precious than its possession.

PART V

CONCLUDING REMARKS:
EINSTEIN THE PHILOSOPHER-SCIENTIST

Einstein's primary objectives were all in the realm of physics. But
he saw that certain physical problems could not be solved unless the
solutions were preceded by a logical analysis of the fundamentals of
space and time, and he saw that this analysis, in turn, presupposed a
philosophic readjustment of a certain familiar conception of knowledge.

—Hans Reichenbach in Schilpp, *Albert Einstein:*
Philosopher-Scientist, p. 290

AMONG THE PROTAGONISTS OF SCHILPP'S LIBRARY OF LIVING PHILOSOPHERS SERIES,
Einstein was the only one not primarily known as a philosopher but rather as a scientist.
In his preface, Schilpp does not waste any words in justifying this decision: "The contents
of this volume speak eloquently enough for the inclusion of *Albert Einstein: Philosopher-
Scientist* in this series without any superfluous words from the editor."

Nor in his *Autobiographical Notes* does Einstein himself feel any need to discuss the
context in which his autobiography appears. He did not think in terms of disciplinary
boundaries. But the lack of understanding he felt for his position in the physics community may have acted as a further motivation to direct his concerns to a philosophical
audience. He certainly saw a lack of epistemological awareness among the community of
physicists as one reason for its indifference and opposition with regard to his program of
a unified field theory. About two years after writing his autobiography, he conveyed in a
letter to Solovine the same sentiments, referring specifically to his search for a unified
field theory: "The unified field theory is now complete. But it is so difficult to employ
mathematically that, despite all efforts, I have been unable to somehow verify it. This
situation will last for many more years, mainly because physicists have no understanding

of logical and philosophical arguments."[1] In this sense, Einstein's research program and his participation in a philosophical project are intimately connected, and the opportunity offered by Schilpp to present himself as a philosopher-scientist perfectly matched his own concerns and ambition.

This may sound, however, too much as if there had been a strategic plan. As with many things, the beginning of the productive collaboration between Schilpp and Einstein was a chance encounter. Schilpp was encouraged by Philipp Frank, Einstein's successor to the physics chair in Prague and himself a philosopher-scientist, to ask Einstein for a half-hour meeting to discuss a specific topic. The meeting took place in December 1940 in Princeton and lasted for an hour and a half. We know what was discussed from the subsequent correspondence between the two. Schilpp gave his host the first volume of the Living Philosophers series, dedicated to John Dewey, the famous American pragmatist. He explained the project, and Einstein was apparently impressed.

Two days later, Schilpp wrote to Einstein thanking him for his positive response to the idea of the series and specifically for his willingness to cooperate on a volume dedicated to Bertrand Russell, the British mathematician, logician, and philosopher. He informed Einstein that he had already written to the publishers of Russell's books asking them to send copies of the books to Einstein. He also asked Einstein to write one or two sentences about his general opinion of the LLP series. On 12 January 1941, Schilpp asked Einstein specifically to contribute something related to Russell's epistemology. On 18 July, Einstein responded that he had read thoroughly and with interest Russell's books that Schilpp had sent him. The purely logical work of Russell was not directly close to his field, he remarked, but he promised to write something when he found the "inner peace" to do so. In November 1942, Einstein received a reminder from Schilpp, also telling him how happy Russell was to hear that Einstein promised to contribute to his volume. Einstein confirmed the agreed title: "Russell's Epistemology."

Einstein's first encounter with Russell's writings left a lasting influence. Years later, he was still fascinated with Bertrand Russell. In a letter to his friend Michele Besso, sent a few months after the death of his sister Maja, Einstein wrote: "I miss her very much. During the years of her illness we read a good part of the best books of all times together. But her absolute favorite was Bertrand Russell—mine too, for that matter."[2]

On 26 February 1943, he submitted his brief text under the title "Remarks on Bertrand Russell's Theory of Knowledge." He obviously struggled with this text, as is evident from the first paragraph and from the letter to Schilpp in which he confessed that he had cursed his original promise several times. Einstein blamed these difficulties on his poor experience in philosophy. Nevertheless, Schilpp was very pleased with what he received. He found it of great scientific and philosophical meaning, in particular because Einstein did not agree with the then prevailing tendency of positivistic empiricism but rather favored what Schilpp described as a "meaningful metaphysics." He showed himself convinced that taking this position would "make history." In one way, it certainly did: as we have discussed in the introduction, this exchange paved the way for Einstein's inclusion in the series as a "philosopher-scientist" three years later.

There can be no doubt that he merited this designation. Einstein's scientific biography is characterized by his deep interest in philosophical issues, starting from his early fascination with Arthur Schopenhauer's writings, his attendance at philosophical lectures on Immanuel Kant at the Polytechnical School in Zurich, and his readings with friends from the Akademie Olympia during his time as a patent clerk in Bern in the early 1900s. The readings that most impressed him at the time still figure prominently in the *Notes*, in particular David Hume and Ernst Mach. Throughout his life, Einstein stressed their importance for his thinking, as he first prominently articulated in 1916: "I know at least that Hume and Mach have helped me a lot, both directly and indirectly."[3]

Einstein's early readings of philosophers such as Hume or philosopher-scientists such as Mach and Poincaré had made him aware of the delicate relation between fundamental concepts such as space and time and experience. In creating special relativity, Hume's empiricism and Poincaré's conventionalism had been helpful because they encouraged Einstein to ascribe new notions of space and time to coordinates that in Lorentz's electrodynamics had no direct physical interpretation. His philosophical awareness was, however, much more than some kind of background knowledge enabling him to address concrete physical problems with greater epistemological sensibility than his contemporaries. Rather, the very development of his theories of relativity made it necessary for Einstein to probe this reflective competence and to engage himself in philosophical debates in order to resolve foundational ambiguities of the emerging theory of general relativity.

Einstein's interest in philosophy was nothing aloof from his actual work in physics. For this reason alone, the hyphenated title "philosopher-scientist" might have pleased him. Indeed, his scientific breakthroughs are intimately connected with his epistemological reflections, in particular on the nature and origin of fundamental concepts such as space, time, and causality. All the more, at the time he was creating general relativity, he appreciated the efforts of a younger generation of philosophers who in turn deeply immersed themselves in the new physical theories, helping their emergence by accompanying them with epistemological reflections informed by a technical understanding of the science. Prominent examples are Moritz Schlick and Hans Reichenbach, two of his main philosophical interlocutors in the formative period of general relativity. Einstein's work, but also his active participation in philosophical discussions in the 1910s and 1920s acted as an important stimulus for the development of contemporary philosophy.

Given the impact of philosophical thought on the genesis and formative years of general relativity, it is little wonder that Einstein felt the need to reconsider the various philosophical positions that had marked his pathway. That was one of the reasons for the intensity with which he interacted with philosophers. Another was the hope to learn himself from what he had achieved and to extract heuristic clues for his own future pathway. A third motive was the attempt to defend the theory and to resolve some of the puzzles it had created.

If Moritz Schlick had not been assassinated about ten years earlier, he would have been most certainly one of the main contributors to Schilpp's Einstein volume. Now, the representation of the Vienna (Berlin) Circle was left to Hans Reichenbach. His remarks and Einstein's response are an extension of their earlier interactions, which started with their discussions of Kantian and neo-Kantian philosophy, following the publication of Reichenbach's book *Relativitätstheorie und Erkenntnis a priori* (*The Theory of Relativity and A Priori Knowledge*). Reichenbach had asked Einstein's permission to dedicate the book to him: "By placing your name at the head of the text, I would like to express how greatly philosophy in particular is indebted to you. I know very well that very few among tenured philosophers have the faintest idea that your theory is a philosophical feat and that your physical conceptions contain more philosophy than all the multi-volume works by the epigones of the great Kant. Do, therefore, please allow me to express these thanks to you with this attempt to free the profound insights of Kantian philosophy from its contemporary trappings and to combine it with your discoveries within a single system."[4] Einstein responded: "I am really very pleased that you want to dedicate your excellent brochure to me, but even more so that you give me such high marks as a lecturer and thinker. The value of the th. of rel. for philosophy seems to me to be that it exposed the dubiousness of certain concepts that even in philosophy were recognized as small change. Concepts are simply empty when they stop being firmly linked to experiences. They resemble upstarts who are ashamed of their origins and want to disown them."[5]

As the historian Klaus Hentschel has discussed in detail, in this phase of his thinking, Reichenbach still saw his position as a reinterpretation of Kant's belief in the necessity of certain a priori forms of thinking.[6] But in contrast to Kant, he emphasized the historically changing character of these forms of thinking and did not subscribe to their apodictic character.

Einstein took a very critical stance not only with regard to Kantian philosophy but also with respect to any such attempt to distinguish within a theory elements that are closer to reality than others. Yet there is one element of Kant's doctrine that Einstein came to understand as truly valuable, as discussed in the exchange of remarks between Einstein and Reichenbach and between Einstein and Margenau (see Part III).

In the formative years of relativity, Einstein increasingly distanced himself from the kind of radical empiricism and positivism that had earlier fascinated him in Ernst Mach. In a seminal paper from 1936 on epistemology, "Physics and Reality," Einstein asserts: "The supreme task of the physicist is to seek those most universal laws from which, by pure deduction, the worldview may be achieved."[7] The emphasis here is on "pure deduction." Einstein would not have made such a statement in the earlier phase of his scientific development. It marks a departure from Mach's and Hume's empiricism. An even sharper departure from that doctrine is expressed in Einstein's recollection about the lesson he learned in the process of developing his general theory of relativity: "I have learned something else from the theory of gravitation: no collection of empirical facts however comprehensive can ever lead to the setting up of such complicated equations. . . . [They] can be found only through the discovery of a logically simple mathematical condition that determines the equations completely or almost completely. Once one has obtained those sufficiently strong formal conditions, one requires only little knowledge of facts for the construction of the theory" (*Notes*, p. 85 [pp. 181–182]).

The departure from strict empiricism is marked, as Einstein emphasizes, by the role of free imagination as a constitutive element of science as well as by the idea that only a theory as a whole can meaningfully be compared with empirical evidence. The latter conviction resonates with the conventionalist position that philosopher-scientists such as Henri Poincaré and, in particular, Pierre Duhem had taken. But it would hardly do justice to Einstein as a philosopher-scientist to try to fit him into one of the numerous drawers that historians of philosophy have used for classifying an epistemological position. In his response to the remarks by Lenzen and Northrop, Einstein, obviously referring to himself, explains that a scientist is different from a philosopher.[8] The scientist cannot rigidly adhere to a particular epistemological position. On different occasions he may sound like a realist, an idealist, a positivist, and even a Platonist. This may strike a systematic epistemologist as "epistemological opportunism." However, the historian and philosopher of science Don Howard remarks on this statement: "When viewed in its proper historical setting, it emerges as an original synthesis of profound and coherent philosophy of science that is of continuing relevance today, the unifying thread of which is, from early to late, the assimilation of Duhem's holistic version of conventionalism."[9]

Einstein thought even more freely than that, not being confined to science and philosophy. He was equally interested in the history of science and the psychology of human thinking. In these fields as well, he had found important interlocutors, such as the founder of Gestalt psychology, Max Wertheimer, a refugee from Germany to the United States, like himself.[10]

By the end of the 1920s, some of Einstein's main philosophical interlocutors, on the other hand, had lost their fervent interest in the living practice and detailed contents of science, looking instead for a foundation to their approaches in an analysis of language, logic, and scientific method. Historians of philosophy speak of a "linguistic turn" associated in particular with the influential writings of Bertrand Russell and Ludwig Wittgenstein, which shaped much of the philosophical discourse in the remainder of the twentieth century.[11] In this context, it seems not unlikely that Schilpp's invitation to Einstein to comment on Russell acted as a catalyst for the creation of the Einstein volume in the Living Philosophers series. They in fact agreed that there must be more to a philosophy of science than the contemporary trend toward a logical empiricism. The fortunate moment in which philosophers had attempted to penetrate science, while scientists felt the need to bother with epistemological reflections in solving their scientific problems had largely passed, at least for the mainstreams of physics and philosophy. In a volume emphasizing the intimate connection between fundamental science and conceptual thinking, Einstein may have also hoped to preserve this unique legacy.

By the time Einstein wrote his *Autobiographical Notes*, the quest for a unified worldview was no longer, as in the 1920s and 1930s, on the agenda of scientists and philosophers. Einstein remained essentially alone in pursuing this goal. He had become a lone traveler not only in this respect. His position both within contemporary physics and philosophy was that of an outsider resisting the main stream of quantum mechanics in one case and logical positivism in the other. With his death, the attempt to develop a worldview based on scientific inquiry and epistemological reflection was marginalized. To some extent this was even true for general relativity itself. In 1955, the first international conference on relativity took place in Bern. This event launched the renaissance of general relativity, bringing it to the stage of contemporary physics.[12] However, this was not a renaissance of the quest for a unified worldview.

NOTES

1. Einstein to Solovine, 12 February 1951, AEA 21-277.
2. Einstein to Besso, 12 December 1951, AEA 7-401.
3. Albert Einstein, "Ernst Mach," *Physikalische Zeitschrift* 17 (1916): 101–104; reprinted in CPAE vol. 6, Doc. 29, p. 143.
4. Reichenbach to Einstein, 15 June 1920, CPAE vol. 10, Doc. 57.
5. Einstein to Reichenbach, 30 June 1920, CPAE vol. 10, Doc. 66.
6. Klaus Hentschel, *Interpretationen und Fehlinterpretationen der speziellen und allgemeinen Relativitaetstheorie durch Zeitgenossen Albert Einstein* (Basel: Birkhauser, 1990), sec. 4.1.

7. Einstein, "Physics and Reality" (1936), reprinted in *Ideas and Opinions: Based on "Mein Weltbild,"* ed. Carl Seelig (New York: Bonanza Books, 1954), 290–323.

8. Einstein's "Reply to Criticisms," in Schilpp, *Albert Einstein: Philosopher-Scientist*, pp. 683–684.

9. Don Howard, "Einstein and the Development of Twentieth-Century Philosophy of Science," in *The Cambridge Companion to Einstein*, ed. Michel Janssen and Christoph Lehner (Cambridge: Cambridge University Press, 2014), 375.

10. For a discussion of this point, see Hanoch Gutfreund and Jürgen Renn, *The Formative Years of Relativity: The History and Meaning of Einstein's Princeton Lecture*s (Princeton, NJ: Princeton University Press, 2017), 118.

11. Fynn Ole Engler and Jürgen Renn, *Gespaltene Vernunft: Vom Ende eines Dialogs zwischen Wissenschaft und Philosophie* (Berlin: Mattes & Seitz, 2018).

12. Alexander Blum, Roberto Lalli, and Jürgen Renn, "The Reinvention of General Relativity: A Historiographical Framework for Assessing One Hundred Years of Curved Space-time," *Isis* 106, no. 3 (2015): 598–620; "The Renaissance of General Relativity: How and Why It Happened," *Ann. Phys.* 528, no. 5 (2016): 344–349; (guest editors), "The Renaissance of Einstein's Theory of Gravitation (special issue of *EPJ H*)," *EPJ H* 42 (2017); "Gravitational Waves and the Long Relativity Revolution," *Nature Astronomy* 2(2018): 534–543.

PART VI

REPRINT OF THE ENGLISH TRANSLATION
OF *AUTOBIOGRAPHICAL NOTES*

AUTOBIOGRAPHICAL NOTES[1]

HERE I SIT IN ORDER TO WRITE, AT THE AGE OF SIXTY-SEVEN, SOMETHING LIKE MY OWN obituary. I am doing this not merely because Dr. Schilpp has persuaded me to do it, but because I do, in fact, believe that it is a good thing to show those who are striving alongside of us how our own striving and searching appears in retrospect. After some reflection, I felt how imperfect any such attempt is bound to be. For, however brief and limited one's working life may be, and however predominant may be the way of error, the exposition of that which is worthy of communication does nonetheless not come easy—today's person of sixty-seven is by no means the same as was the one of fifty, of thirty, or of twenty. Every reminiscence is colored by one's present state, hence by a deceptive point of view. This consideration could easily deter one. Nevertheless much can be gathered out of one's own experience that is not open to another consciousness.

When I was a fairly precocious young man I became thoroughly impressed with the futility of the hopes and strivings that chase most men restlessly through life. Moreover, I soon discovered the cruelty of that chase, which in those years was much more carefully covered up by hypocrisy and glittering words than is the case today. By the mere existence of his stomach everyone was condemned to participate in that chase. The stomach might well be satisfied by such participation, but not man insofar as he is a thinking and feeling being. As the first way out there was religion, which is implanted into every child by way of the traditional education-machine. Thus I came—though the child of entirely irreligious (Jewish) parents—to a deep religiousness, which, however, reached an abrupt end at the age of twelve. Through the reading of popular scientific books I soon reached the conviction that much in the stories of the Bible could not be true. The consequence was a positively fanatic [orgy of] freethinking coupled with the impression that youth is intentionally being deceived by the state through lies; it was a crushing impression. Mistrust of every kind of authority grew out of this experience, a skeptical attitude toward the convictions that were alive in any specific social environment—an attitude that has never again left me, even though, later on, it has been tempered by a better insight into the causal connections.

It is quite clear to me that the religious paradise of youth, which was thus lost, was a first attempt to free myself from the chains of the "merely personal," from an existence dominated by wishes, hopes, and primitive feelings. Out yonder there was this huge world, which exists independently of us human beings and which stands before us like

1   Translated from the original German manuscript by Paul Arthur Schilpp and revised with the help of Professor Peter Bergmann of Syracuse University.

a great, eternal riddle, at least partially accessible to our inspection and thinking. The contemplation of this world beckoned as a liberation, and I soon noticed that many a man whom I had learned to esteem and to admire had found inner freedom and security in its pursuit. The mental grasp of this extra-personal world within the frame of our capabilities presented itself to my mind, half consciously, half unconsciously, as a supreme goal. Similarly motivated men of the present and of the past, as well as the insights they had achieved, were the friends who could not be lost. The road to this paradise was not as comfortable and alluring as the road to the religious paradise; but it has shown itself reliable, and I have never regretted having chosen it.

What I have said here is true only in a certain sense, just as a drawing consisting of a few strokes can do justice to a complicated object, full of perplexing details, only in a very limited sense. If an individual enjoys well-ordered thoughts, it is quite possible that this side of his nature may grow more pronounced at the cost of other sides and thus may determine his mentality in increasing degree. In this case it may well be that such an individual sees in retrospect a uniformly systematic development, whereas the actual experience takes place in kaleidoscopic particular situations. The great variety of the external situations and the narrowness of the momentary content of consciousness bring about a sort of atomizing of the life of every human being. In a man of my type, the turning point of the development lies in the fact that gradually the major interest disengages itself to a far-reaching degree from the momentary and the merely personal and turns toward the striving for a conceptual grasp of things. Looked at from this point of view, the above schematic remarks contain as much truth as can be stated with such brevity.

What, precisely, is "thinking"? When, on the reception of sense impressions, memory pictures emerge, this is not yet "thinking." And when such pictures form sequences, each member of which calls forth another, this too is not yet "thinking." When, however, a certain picture turns up in many such sequences, then—precisely by such return—it becomes an organizing element for such sequences, in that it connects sequences in themselves unrelated to each other. Such an element becomes a tool, a concept. I think that the transition from free association or "dreaming" to thinking is characterized by the more or less preeminent role played by the "concept." It is by no means necessary that a concept be tied to a sensorily cognizable and reproducible sign (word); but when this is the case, then thinking becomes thereby capable of being communicated.

With what right—the reader will ask—does this man operate so carelessly and primitively with ideas in such a problematic realm without making even the least effort to prove anything? My defense: all our thinking is of this nature of free play with concepts; the justification for this play lies in the degree of comprehension of our sensations that we are able to achieve with its aid. The concept of "truth" can not yet be applied to such a structure; to my thinking this concept becomes applicable only when a far-reaching agreement (*convention*) concerning the elements and rules of the game is already at hand.

I have no doubt but that our thinking goes on for the most part without use of signs (words) and beyond that to a considerable degree unconsciously. For how, otherwise, should it happen that sometimes we "wonder" quite spontaneously about some experience? This "wondering" appears to occur when an experience comes into conflict with a world of concepts already sufficiently fixed within us. Whenever such a conflict is experienced sharply and intensively it reacts back upon our world of thought in a decisive way.

The development of this world of thought is in a certain sense a continuous flight from "wonder."

A wonder of this kind I experienced as a child of four or five years when my father showed me a compass. That this needle behaved in such a determined way did not at all fit into the kind of occurrences that could find a place in the unconscious world of concepts (efficacy produced by direct "touch"). I can still remember—or at least believe I can remember—that this experience made a deep and lasting impression upon me. Something deeply hidden had to be behind things. What man sees before him from infancy causes no reaction of this kind; he is not surprised by the falling of bodies, by wind and rain, nor by the moon, nor by the fact that the moon does not fall down, nor by the differences between living and nonliving matter.

At the age of twelve I experienced a second wonder of a totally different nature—in a little book dealing with Euclidean plane geometry, which came into my hands at the beginning of a school year. Here were assertions, as for example the intersection of the three altitudes of a triangle at one point, that—though by no means evident—could nevertheless be proved with such certainty that any doubt appeared to be out of the question. This lucidity and certainty made an indescribable impression upon me. That the axioms had to be accepted unproved did not disturb me. In any case it was quite sufficient for me if I could base proofs on propositions whose validity appeared to me beyond doubt. For example, I remember that an uncle told me about the Pythagorean theorem before the holy geometry booklet had come into my hands. After much effort I succeeded in "proving" this theorem on the basis of the similarity of triangles; in doing so it seemed to me "evident" that the relations of the sides of the right-angled triangles would have to be completely determined by one of the acute angles. Only whatever did not in similar fashion seem to be "evident" appeared to me to be in need of any proof at all. Also, the objects with which geometry is concerned seemed to be of no different type from the objects of sensory perception, "which can be seen and touched." This primitive conception, which probably also lies at the bottom of the well-known Kantian inquiry concerning the possibility of "synthetic judgments *a priori*," rests obviously upon the fact that the relation of geometrical concepts to objects of direct experience (rigid rod, finite interval, etc.) was unconsciously present.

If thus it appeared that it was possible to achieve certain knowledge of the objects of experience by means of pure thinking, this "wonder" rested upon an error. Nevertheless, for anyone who experiences it for the first time, it is marvelous enough that man is capable at all of reaching such a degree of certainty and purity in pure thinking as the Greeks showed us for the first time to be possible in geometry.

Now that I have allowed myself to be carried away sufficiently to interrupt my barely started obituary, I shall not hesitate to state here in a few sentences my epistemological credo, although in what precedes something has already incidentally been said about this. This credo actually evolved only much later and very slowly and does not correspond to the point of view I held in younger years.

I see on the one side the totality of sense experiences and, on the other, the totality of the concepts and propositions that are laid down in books. The relations between the concepts and propositions among themselves are of a logical nature, and the business of logical thinking is strictly limited to the achievement of the connection between concepts and propositions among themselves according to firmly laid down rules, which are the concern of logic. The concepts and propositions get "meaning," or "content," only through their connection with sense experiences. The connection of the latter with the former is purely intuitive, not itself of a logical nature. The degree of certainty with which this connection, or intuitive linkage, can be undertaken, and nothing else, differentiates empty fantasy from scientific "truth." The system of concepts is a creation of man, together with the rules of syntax, which constitute the structure of the conceptual systems. Although the conceptual systems are logically entirely arbitrary, they are restricted by the aim of permitting the most nearly possible certain (intuitive) and complete coordination with the totality of sense experiences; secondly they aim at the greatest possible sparsity of their logically independent elements (basic concepts and axioms), i.e., their undefined concepts and underived [postulated] propositions.

A proposition is correct if, within a logical system, it is deduced according to the accepted logical rules. A system has truth-content according to the certainty and completeness of its possibility of coordination with the totality of experience. A correct proposition borrows its "truth" from the truth-content of the system to which it belongs.

A remark as to the historical development. Hume saw clearly that certain concepts, as for example that of causality, cannot be deduced from the material of experience by logical methods. Kant, thoroughly convinced of the indispensability of certain concepts, took them—just as they are selected—to be the necessary premises of any kind of thinking and distinguished them from concepts of empirical origin. I am convinced, however, that this distinction is erroneous or, at any rate, that it does not do justice to the problem in a natural way. All concepts, even those closest to experience, are from the point of view of logic freely chosen posits, just as is the concept of causality, which was the point of departure for this inquiry in the first place.

And now back to the obituary. At the age of twelve through sixteen I familiarized myself with the elements of mathematics, including the principles of differential and integral calculus. In doing so I had the good fortune of encountering books that were not too particular regarding logical rigor, but that permitted the principal ideas to stand out clearly. This occupation was, on the whole, truly fascinating; there were peaks whose impression could easily compete with that of elementary geometry—the basic idea of analytical geometry, the infinite series, the concepts of derivative and integral. I also had the good fortune of getting to know the essential results and methods of the entire field of the natural sciences in an excellent popular exposition, which limited itself almost throughout to qualitative aspects (Bernstein's *Popular Books on Natural Science*, a work of five or six volumes), a work that I read with breathless attention. I had also already studied some theoretical physics when, at the age of seventeen, I entered the Polytechnic Institute of Zürich as a student of mathematics and physics.

There I had excellent teachers (for example, Hurwitz, Minkowski), so that I should have been able to obtain a mathematical training in depth. I worked most of the time in the physical laboratory, however, fascinated by the direct contact with experience. The balance of the time I used, in the main, in order to study at home the works of Kirchhoff, Helmholtz, Hertz, etc. The fact that I neglected mathematics to a certain extent had its cause not merely in my stronger interest in the natural sciences than in mathematics but also in the following peculiar experience. I saw that mathematics was split up into numerous specialties, each of which could easily absorb the short lifetime granted to us. Consequently, I saw myself in the position of Buridan's ass, which was unable to decide upon any particular bundle of hay. Presumably this was because my intuition was not strong enough in the field of mathematics to differentiate clearly the fundamentally important, that which is really basic, from the rest of the more or less dispensable erudition. Also, my interest in the study of nature was no doubt stronger; and it was not clear to me as a young student that access to a more profound knowledge of the basic principles of physics depends on the most intricate mathematical methods. This dawned upon me only gradually after years of independent scientific work. True enough, physics also was divided into separate fields, each of which was capable of devouring a short lifetime of work without having satisfied the hunger for deeper knowledge. The mass of insufficiently connected experimental data was overwhelming here also. In this field, however, I soon learned to scent out that which might lead to fundamentals and to turn aside from everything else, from the multitude of things that clutter up the mind and divert it from the essentials. The hitch in this was, of course, that one had to cram all this stuff into one's mind for the examinations, whether one liked it or not. This coercion had such a deterring effect [upon me] that, after I had passed the final examination, I found the consideration of any scientific problems distasteful to me for an entire year. Yet I must say that in Switzerland we had to suffer far less under such coercion, which smothers every truly scientific impulse, than is the case in many another locality. There were altogether only two examinations; aside from these, one could just about do as one pleased. This was especially the case if one had a friend, as did I, who attended the lectures regularly and who worked over their content conscientiously. This gave one freedom in the choice of pursuits until a few months before the examination, a freedom I enjoyed to a great extent, and I have gladly taken into the bargain the resulting guilty conscience as by far the lesser evil. It is, in fact, nothing short of a miracle that the modern methods of instruction have not yet entirely strangled the holy curiosity of inquiry; for this delicate little plant, aside from stimulation, stands mainly in need of freedom; without this it goes to wrack and ruin without fail. It is a very grave mistake to think that the enjoyment of seeing and searching can be promoted by means of coercion and a sense of duty. To the contrary, I believe that it would be possible to rob even a healthy beast of prey of its voraciousness if it were possible, with the aid of a whip, to force the beast to take food continuously even when not hungry, especially if the food handed out under such coercion were to be selected accordingly.

Now to the field of physics as it presented itself at that time. In spite of great productivity in particulars, dogmatic rigidity prevailed in matters of principle: In the beginning (if there was such a thing), God created Newton's laws of motion together with the necessary masses and forces. This is all; everything beyond this follows from the development of appropriate mathematical methods by means of deduction. What the nineteenth century achieved on the strength of this basis, especially through the application of partial differential equations, was bound to arouse the admiration of every receptive person. Newton was probably first to reveal, in his theory of the propagation of sound, the efficacy of partial differential equations. Euler had already created the foundation of hydrodynamics. But the more sophisticated development of the mechanics of discrete masses, as the basis of all physics, was the achievement of the nineteenth century. What made the greatest impression upon the student, however, was not so much the technical development of mechanics or the solution of complicated problems as the achievements of mechanics in areas that apparently had nothing to do with mechanics: the mechanical theory of light, which conceived of light as the wave motion of a quasi-rigid elastic ether; and above all the kinetic theory of gases: the independence of the specific heat of monatomic gases from the atomic weight, the derivation of the equation of the state of a gas and its relation to the specific heat, the kinetic theory of the dissociation of gases, and above all the quantitative relationship between viscosity, heat conduction, and diffusion of gases, which also furnished the absolute magnitude of the atom. These results supported at the same time mechanics as the foundation of physics and of the atomic hypothesis, which latter was already firmly rooted in chemistry. In chemistry, however, only the ratios of the atomic masses played any role, not their absolute magnitudes, so that atomic theory could be viewed more as a visualizing symbol than as knowledge concerning the actual composition of matter. Apart from this it was also of profound interest that the statistical theory of classical mechanics was able to deduce the basic laws of thermodynamics, something in essence already accomplished by Boltzmann.

We must not be surprised, therefore, that, so to speak, all physicists of the previous century saw in classical mechanics a firm and definitive foundation for all physics, indeed for the whole of natural science, and that they never grew tired in their attempts to base Maxwell's theory of electromagnetism, which, in the meantime, was slowly beginning to win out, upon mechanics as well. Even Maxwell and H. Hertz, who in retrospect are properly recognized as those who shook the faith in mechanics as the final basis of all physical thinking, in their conscious thinking consistently held fast to mechanics as the confirmed basis of physics. It was Ernst Mach who, in his *History of Mechanics*, upset this dogmatic faith; this book exercised a profound influence upon me in this regard while I was a student. 1 see Mach's greatness in his incorruptible skepticism and independence; in my younger years, however, Mach's epistemological position also influenced me very greatly, a position that today appears to me to be essentially untenable. For he did not place in the correct light the essentially constructive and speculative nature of all thinking and more especially of scientific thinking; in consequence, he condemned theory precisely at those points where its constructive-speculative character comes to light unmistakably, such as in the kinetic theory of atoms.

Before I enter upon a critique of mechanics as the foundation of physics, something general will have to be said first about the points of view from which physical theories may be analyzed critically at all. The first point of view is obvious: the theory must not contradict empirical facts. However evident this demand may in the first place appear, its application turns out to be quite delicate. For it is often, perhaps even always, possible to retain a general theoretical foundation by adapting it to the facts by means of artificial additional assumptions. In any case, however, this first point of view is concerned with the confirmation of the theoretical foundation by the available empirical facts.

The second point of view is not concerned with the relationship to the observations but with the premises of the theory itself, with what may briefly but vaguely be characterized as the "naturalness" or "logical simplicity" of the premises (the basic concepts and the relations between these). This point of view, whose exact formulation meets with great difficulties, has played an important role in the selection and evaluation of theories from time immemorial. The problem here is not simply one of a kind of enumeration of the logically independent premises (if anything like this were at all possible without ambiguity), but one of a kind of reciprocal weighing of incommensurable qualities. Furthermore, among theories with equally "simple" foundations, that one is to be taken as superior which most sharply delimits the otherwise feasible qualities of systems (i.e., contains the most specific claims). Of the "scope" of theories I need not speak here, inasmuch as we are confining ourselves to such theories as have for their object the *totality* of all physical phenomena. The second point of view may briefly be characterized as concerned with the "inner perfection" of the theory, whereas the first point of view refers to the "external confirmation." The following I reckon as also belonging to the "inner perfection" of a theory: We prize a theory more highly if, from the logical standpoint, it does not involve an arbitrary choice among theories that are equivalent and possess analogous structures.

I shall not attempt to excuse the lack of precision of the assertions contained in the last two paragraphs on the grounds of insufficient space at my disposal; I must confess herewith that I cannot at this point, and perhaps not at all, replace these hints by more precise definitions. I believe, however, that a sharper formulation would be possible. In any case it turns out that among the "oracles" there usually is agreement in judging the "inner perfection" of the theories and even more so concerning the degree of "external confirmation."

And now to the critique of mechanics as the basis of physics.

From the first point of view (confirmation by experiment) the incorporation of wave optics into the mechanical picture of the world was bound to arouse serious misgivings. If light was to be interpreted as undulatory motion in an elastic body (ether), this had to be a medium that permeates everything, because of the transversality of the light waves, in the main resembling a solid body, yet incompressible, so that longitudinal waves did not exist. This ether had to lead a ghostly existence alongside the rest of matter, inasmuch as it seemed to offer no resistance whatever to the motion of "ponderable" bodies. In order to explain the indices of refraction of transparent bodies as well as the processes of emission and absorption of radiation, one would have had to assume complicated interactions between the two types of matter, something that was not even seriously tried, let alone achieved.

Furthermore, the electromagnetic forces necessitated the introduction of electric masses that, although they had no noticeable inertia, yet interacted with each other and whose interaction was, moreover, in contrast to the force of gravitation, of a polar type.

What eventually made the physicists abandon, after hesitating a long time, their faith in the possibility that all physics could be founded upon Newton's mechanics, was the electrodynamics of Faraday and Maxwell. For this theory and its confirmation by Hertz's experiments showed that there are electromagnetic phenomena that by their very nature are detached from all ponderable matter—namely the waves in empty space that consist of electromagnetic "fields." If mechanics was to be maintained as the foundation of physics, Maxwell's equations had to be interpreted mechanically. This was zealously but fruitlessly attempted, whereas the equations themselves turned out to be increasingly fruitful. One got used to operating with these fields as independent substances without finding it necessary to account for their mechanical nature; thus mechanics as the basis of physics was being abandoned, almost imperceptibly, because its adaptation to the facts presented itself finally as a hopeless task. Since then, there exist two types of conceptual elements: on the one hand, material points with forces at a distance between them and, on the other hand, the continuous field. We are at an intermediate state of physics without a uniform basis for the whole, a state that—although unsatisfactory—is far from having been overcome.

Now for a few remarks concerning the critique of mechanics as the foundation of physics from the second, the "interior," point of view. In today's state of science, i.e., after the abandonment of the mechanical foundation, such a critique retains only a methodological relevance. But such a critique is well suited to show the type of argumentation that, in the selection of theories in the future, will have to play an ever greater role the more the basic concepts and axioms are removed from what is directly observable, so that the confrontation of the implications of theory by the facts becomes constantly more difficult and more drawn out. First in line to be mentioned is Mach's argument, which, incidentally, had already been clearly recognized by Newton (bucket experiment). From the standpoint of purely geometrical description, all "rigid" coordinate systems are logically equivalent. The equations of mechanics (for example the law of inertia) claim validity only when referred to a specific class of such systems, i.e., the "inertial systems." In this connection the coordinate system as a material object is without any significance. Hence to justify the need for this specific choice one must search for something that exists beyond the objects (masses, distances) with which the theory deals. For this reason "absolute space" as originally determinative was quite explicitly introduced by Newton as the omnipresent active participant in all mechanical events; by "absolute" he obviously means: uninfluenced by the masses and by their motion. What makes this state of affairs appear particularly ugly is the fact that there are supposed to be infinitely many inertial systems, relative to each other in uniform and irrotational translation, which are supposed to be distinguished among all other rigid systems.

Mach conjectures that in a truly reasonable theory inertia would have to depend upon the interaction of the masses, precisely as was true for Newton's other forces, a conception that for a long time I considered in principle the correct one. It presupposes implicitly, however, that the basic theory should be of the general type of Newton's mechanics: masses and their interaction as the original concepts. Such an attempt at a resolution does not fit into a consistent field theory, as will be immediately recognized.

How sound, however, Mach's critique is in essence can be seen particularly clearly from the following analogy. Let us imagine people who construct a mechanics, who know only a very small part of the earth's surface and who also cannot see any stars. They will be inclined to ascribe special physical attributes to the vertical dimension of space (direction of the acceleration of falling bodies) and, on the ground of such a conceptual basis, will offer reasons that the earth is in most places horizontal. They might not let themselves be influenced by the argument that in its geometrical properties space is isotropic and that it is therefore unsatisfactory to postulate basic physical laws according to which there is to be a preferential direction; they will probably be inclined (analogously to Newton) to assert the absoluteness of the vertical, as proved by experience, as something with which one simply would have to come to terms. The preference given to the vertical over all other spatial directions is precisely analogous to the preference given to inertial systems over other rigid coordinate systems.

Now to [a consideration of] other arguments that also concern themselves with the inner simplicity, or naturalness, of mechanics. If one accepts the concepts of space (including geometry) and time without critical doubts, then there exists no reason to object to the idea of action at a distance, even though such a concept is unsuited to the ideas one forms on the basis of the raw experience of daily life. However, there is another consideration that makes mechanics, taken as the basis of physics, appear primitive. Essentially there are two laws:

(1) the law of motion
(2) the expression for the force or the potential energy.

The law of motion is precise, although empty as long as the expression for the forces is not given. For postulating the latter, however, there is an enormous degree of arbitrariness, especially if one drops the requirement, which is not very natural in any case, that the forces depend only on the coordinates (and not, for example, on their derivatives with respect to time). Within the framework of that theory alone it is entirely arbitrary that the forces of gravitation (and electricity), which come from one point, are governed by the potential function $(1/r)$. Additional remark: it has long been known that this function is the spherically symmetric solution of the simplest (rotation-invariant) differential equation $\nabla^2 \phi = 0$; it would therefore not have been far-fetched to regard this as a clue that this function was to be considered as resulting from a spatial law, an approach that would have eliminated the arbitrariness in the force law. This is really the first insight that suggests a turning away from the theory of action at a distance, a development that— prepared by Faraday, Maxwell, and Hertz—really begins only later in response to the external pressure of experimental data.

I would also like to mention, as one internal asymmetry of this theory, that the inertial mass that occurs in the law of motion also appears in the law of the gravitational force, but not in the expressions for the other forces. Finally I would like to point to the fact that the division of energy into two essentially different parts, kinetic and potential energy, must be felt to be unnatural; H. Hertz felt this to be so disturbing that, in his very last work, he attempted to free mechanics from the concept of potential energy (i.e., from the concept of force).

Enough of this. Newton, forgive me; you found just about the only way possible in your age for a man of highest reasoning and creative power. The concepts that you created are even today still guiding our thinking in physics, although we now know that they will have to be replaced by others farther removed from the sphere of immediate experience, if we aim at a profounder understanding of relationships.

"Is this supposed to be an obituary?" the astonished reader will likely ask. I would like to reply: essentially yes. For the essential in the being of a man of my type lies precisely in *what* he thinks and *how* he thinks, not in what he does or suffers. Consequently, the obituary can limit itself in the main to the communicating of thoughts that have played a considerable role in my endeavors. A theory is the more impressive the greater the simplicity of its premises, the more different kinds of things it relates, and the more extended its area of applicability. Hence the deep impression that classical thermodynamics made upon me. It is the only physical theory of universal content concerning which I am convinced that, within the framework of the applicability of its basic concepts, it will never be overthrown (for the special attention of those who are skeptics on principle).

The most fascinating subject at the time that I was a student was Maxwell's theory. What made this theory appear revolutionary was the transition from action at a distance to fields as the fundamental variables. The incorporation of optics into the theory of electromagnetism, with its relation of the speed of light to the electric and magnetic absolute system of units as well as the relation of the index of refraction to the dielectric constant, the qualitative relation between the reflection coefficient of a body and its metallic conductivity—it was like a revelation. Aside from the transition to field theory, i.e., the expression of the elementary laws through differential equations, Maxwell needed only one single hypothetical step—the introduction of the electrical displacement current in the vacuum and in the dielectrica and its magnetic effect, an innovation that was almost preordained by the formal properties of the differential equations. In this connection I cannot suppress the remark that the pair Faraday-Maxwell has a most remarkable inner similarity with the pair Galileo-Newton—the former of each pair grasping the relations intuitively, and the second one formulating those relations exactly and applying them quantitatively.

What rendered the insight into the essence of electromagnetic theory so much more difficult at that time was the following peculiar situation. Electric or magnetic "field intensities" and "displacements" were treated as equally elementary variables, empty space as a special instance of a dielectric body. *Matter* appeared as the bearer of the field, not *space*. By this it was implied that the carrier of the field should have velocity, and this was naturally to apply to the "vacuum" (ether) also. Hertz's electrodynamics of moving bodies rests entirely upon this fundamental attitude.

It was the great merit of H. A. Lorentz that he brought about a change here in a convincing fashion. In principle a field exists, according to him, only in empty space. Matter—considered to consist of atoms—is the only seat of electric charges; between the material particles there is empty space, the seat of the electromagnetic field, which is produced by the position and velocity of the point charges located on the material particles. Dielectric behavior, conductivity, etc., are determined exclusively by the type of mechanical bindings between the particles that constitute the bodies. The particle charges create the field, which, on the other hand, exerts forces upon the charges of the particles, thus determining the motion of the latter according to Newton's law of motion. If one compares this with Newton's system, the change consists in this: action at a distance is replaced by the field, which also describes the radiation. Gravitation is usually not taken into account because of its relative smallness; its inclusion, however, was always possible by enriching the structure of the field, that is to say, by expanding Maxwell's field laws. The physicist of the present generation regards the point of view achieved by Lorentz as the only possible one; at that time, however, it was a surprising and audacious step, without which the later development would not have been possible.

If one views this phase of the development of theory critically, one is struck by the dualism that lies in the fact that the material point in Newton's sense and the field as continuum are used as elementary concepts side by side. Kinetic energy and field energy appear as essentially different things. This appears all the more unsatisfactory as, according to Maxwell's theory, the magnetic field of a moving electric charge represents inertia. Why not then the *whole* of inertia? Then only field energy would be left, and the particle would be merely a domain containing an especially high density of field energy. In that case one could hope to deduce the concept of the mass point together with the equations of motion of the particles from the field equations—the disturbing dualism would have been removed.

H. A. Lorentz knew this very well. However, Maxwell's equations did not permit the derivation of the equilibrium of the electricity that constitutes a particle. Only different, nonlinear field equations could possibly accomplish such a thing. But no method existed for discovering such field equations without deteriorating into adventurous arbitrariness. In any case, one could believe that it would be possible by and by to find a new and secure foundation for all of physics upon the path so successfully initiated by Faraday and Maxwell.

Accordingly, the revolution begun by the introduction of the field was by no means finished. Then it happened that, around the turn of the century, independently of what we have just been discussing, a second fundamental crisis set in, the seriousness of which was suddenly recognized owing to Max Planck's investigations into heat radiation (1900). The history of this event is all the more remarkable because, at least in its first phase, it was not in any way influenced by any surprising discoveries of an experimental nature.

On thermodynamic grounds Kirchhoff had concluded that the energy density and the spectral composition of radiation in a cavity enclosed by impervious walls of the temperature T, must be independent of the nature of the walls. That is to say, the monochromatic density of radiation ρ is a universal function of the frequency v and of the absolute temperature T. Thus arose the interesting problem of determining this function $\rho(v,T)$. What could theoretically be ascertained about this function? According to Maxwell's theory the radiation had to exert a pressure on the walls, determined by the total energy density. From this Boltzmann concluded, by means of pure thermodynamics, that the entire energy density of the radiation ($\int \rho dv$) is proportional to T^4. In this way he found a theoretical justification of a law that had previously been discovered empirically by Stefan; i.e., in this way he connected this empirical law with the basis of Maxwell's theory. Thereafter, by way of an ingenious thermodynamic consideration, which also made use of Maxwell's theory, W. Wien found that the universal function ρ of the two variables v and T would have to be of the form

$$\rho \approx v^3 f\left(\frac{v}{T}\right),$$

whereby $f(v/T)$ is a universal function of the one variable v/T. It was clear that the theoretical determination of this universal function f was of fundamental importance—this was precisely the task that confronted Planck. Careful measurements had led to a rather precise empirical determination of the function f. Relying on those empirical measurements, he succeeded in the first place in finding a statement that rendered the measurements very well indeed:

$$\rho = \frac{8\pi h v^3}{c^3} \frac{1}{exp(hv/kT)-1}$$

whereby h and k are two universal constants, the first of which led to quantum theory. Because of the denominator, this formula looks a bit queer. Was it possible to derive it theoretically? Planck actually did find a derivation, the imperfections of which remained at first hidden, which latter fact was most fortunate for the development of physics. If this formula was correct, it permitted, with the aid of Maxwell's theory, the calculation of the average energy E of a quasi-monochromatic oscillator within the field of radiation:

$$E = \frac{h\nu}{exp(h\nu/kT) - 1}.$$

Planck preferred to attempt calculating this latter magnitude theoretically. In this effort, thermodynamics, for the time being, no longer proved helpful, and neither did Maxwell's theory. This expression had one aspect that was most encouraging. For high temperatures (with ν fixed) it yielded the expression

$$E = kT.$$

This is the same expression obtained in the kinetic theory of gases for the average energy of a mass point capable of oscillating elastically in one dimension. For in kinetic gas theory one gets

$$E = (R/N)T,$$

where R denotes the gas constant, and N the number of molecules per mole, from which constant one can compute the absolute size of the atom. Equating these two expressions one gets

$$N = R/k.$$

The one constant of Planck's formula consequently furnishes exactly the correct size of the atom. The numerical value agreed satisfactorily with the determinations of N by means of kinetic gas theory, though the latter were not very accurate.

This was a great success, which Planck clearly recognized. But the matter has a serious drawback, which Planck fortunately overlooked at first. For the same considerations demand in fact that the relation $E = kT$ would also have to be valid for low temperatures. In that case, however, it would be all over with Planck's formula and with the constant h. From the existing theory, therefore, the correct conclusion would have been: the average kinetic energy of the oscillator is either given incorrectly by the theory of gases, which would imply a refutation of [statistical] mechanics; or else the average energy of the oscillator follows incorrectly from Maxwell's theory, which would imply a refutation of the latter. Under such circumstances it is most probable that both theories are correct only in the limit, but are otherwise false; this is indeed the situation, as we shall see in what follows. If Planck had drawn this conclusion, he probably would not have made his great discovery, because pure deductive reasoning would have been left without a foundation.

Now back to Planck's reasoning. On the basis of the kinetic theory of gases Boltzmann had discovered that, aside from a constant factor, entropy was equal to the logarithm of the "probability" of the state under consideration. Through this insight he recognized the nature of processes that, within the meaning of thermodynamics, are "irreversible." Seen from the molecular-mechanical point of view, however, all processes are reversible. If one calls a state defined in terms of the molecular theory a microscopically described one, or, more briefly, a micro-state, and a state described in terms of thermodynamics a macro-state, then an immensely large number (Z) of states belong to a macroscopic condition. Z then is a measure of the probability of a chosen macro-state. This idea appears to be of outstanding importance also because its applicability is not limited to a microscopic description on the basis of mechanics. Planck recognized this and applied Boltzmann's principle to a system consisting of very many resonators of the same frequency v. The macroscopic state is given by the total energy of the oscillation of all resonators, a micro-state by the fixation of the (instantaneous) energy of each individual resonator. In order to be able to express the number of micro-states belonging to a macro-state by means of a finite number, he [Planck] divided the total energy into a large but finite number of identical energy elements ξ and asked: in how many ways can these energy elements be divided among the resonators. The logarithm of this number, then, furnishes the entropy and thus (via thermodynamics) the temperature of the system. Planck got his radiation formula if he chose his energy elements ξ to have the magnitude $\xi = hv$. The decisive element in this procedure is that the result depends on taking for ξ a definite finite value, i.e., on not going to the limit $\xi = 0$. This form of reasoning does not make obvious the fact that it contradicts the mechanical and electrodynamic basis upon which the derivation otherwise depends. Actually, however, the derivation presupposes implicitly that energy can be absorbed and emitted by the individual resonator only in "quanta" of magnitude hv, i.e., that the energy of a mechanical structure capable of oscillations as well as the energy of radiation can be transferred only in such quanta—in contradiction to the laws of mechanics and electrodynamics. The contradiction with dynamics was here fundamental; whereas the contradiction with electrodynamics might be less fundamental. For the expression for the density of radiation energy, though *compatible* with Maxwell's equations, is not a necessary consequence of these equations. That this expression furnishes important mean values is shown by the fact that the Stefan-Boltzmann law and Wien's law, which are based on it, are in agreement with experience.

All of this was quite clear to me shortly after the publication of Planck's fundamental work; so that, without having a substitute for classical mechanics, I could nevertheless see to what kind of consequences this law of temperature radiation leads for the photoelectric effect and for other related phenomena of the transformation of radiation energy, as well as for the specific heat of (especially) solid bodies. All my attempts, however, to adapt the theoretical foundation of physics to this [new type of] knowledge failed completely. It was as if the ground had been pulled out from under one, with no firm foundation to be seen anywhere upon which one could have built. That this insecure and contradictory foundation was sufficient to enable a man of Bohr's unique instinct and sensitivity to discover the principal laws of the spectral lines and of the electron shells of the atoms, together with their significance for chemistry, appeared to me as a miracle—and appears to me a miracle even today. This is the highest form of musicality in the sphere of thought.

My own interest in those years was less concerned with the detailed consequences of Planck's results, however important these might be. My main question was: What general conclusions can be drawn from the radiation formula concerning the structure of radiation and even more generally concerning the electromagnetic foundation of physics? Before I take this up, I must briefly mention a number of investigations that relate to the Brownian motion and related objects (fluctuation phenomena) and that in essence rest upon classical molecular mechanics. Not acquainted with the investigations of Boltzmann and Gibbs, which had appeared earlier and actually exhausted the subject, I developed the statistical mechanics and the molecular-kinetic theory of thermodynamics based upon it. My principal aim in this was to find facts that would guarantee as much as possible the existence of atoms of definite finite size. In the midst of this I discovered that, according to atomistic theory, there would have to be a movement of suspended microscopic particles capable of being observed, without knowing that observations concerning the Brownian motion were already long familiar. The simplest derivation rested upon the following consideration. If the molecular-kinetic theory is essentially correct, a suspension of visible particles must possess the same kind of osmotic pressure satisfying the gas laws as a solution of molecules. This osmotic pressure depends upon the actual magnitude of the molecules, i.e., upon the number of molecules in a gram-equivalent. If the density of the suspension is inhomogeneous, the osmotic pressure is inhomogeneous, too, and gives rise to a compensating diffusion, which can be calculated from the known mobility of the particles. This diffusion can, on the other hand, also be considered the result of the random displacement—originally of unknown magnitude—of the suspended particles owing to thermal agitation. By comparing the amounts obtained for the diffusion current from both types of reasoning, one obtains quantitatively the statistical law for those displacements, i.e., the law of the Brownian motion. The agreement of these considerations with experience together with Planck's determination of the true molecular size from the law of radiation (for high temperatures) convinced the skeptics, who were quite numerous at that time (Ostwald, Mach), of the reality of atoms. The hostility of these scholars toward atomic theory can undoubtedly be traced back to their positivistic philosophical attitude. This is an interesting example of the fact that even scholars of audacious spirit and fine instinct can be hindered in the interpretation of facts by philosophical prejudices. The prejudice—which has by no means disappeared—consists in the belief that facts by themselves can and should yield scientific knowledge without free conceptual construction. Such a misconception is possible only because one does not easily become aware of the free choice of such concepts, which, through success and long usage, appear to be immediately connected with the empirical material.

The success of the theory of the Brownian motion showed again conclusively that classical mechanics always led to trustworthy results whenever it was applied to motions in which the higher time derivatives of the velocity are negligible. Upon this recognition a relatively direct method can be based that permits us to learn something concerning the constitution of radiation from Planck's formula. One may argue that in a space filled with radiation a freely moving (vertically to its plane), quasi-monochromatically reflecting mirror would have to go through a kind of Brownian movement, the mean kinetic energy of which equals $\frac{1}{2}$ (R/N) T (R = gas constant for one gram-molecule, N = the number of molecules per mole, T = absolute temperature). If radiation were not subject to local fluctuations, the mirror would gradually come to rest because, owing to its motion, it reflects more radiation on its front than on its reverse side. The mirror, however, must experience certain random fluctuations of the pressure exerted upon it because of the fact that the wave packets, constituting the radiation, interfere with one another. These can be computed from Maxwell's theory. This calculation, then, shows that these pressure variations (especially in the case of small radiation densities) are by no means sufficient to impart to the mirror the average kinetic energy $\frac{1}{2}$ (R/N) T. In order to get this result one has to assume rather that there exists a second type of pressure variations, not derivable from Maxwell's theory, corresponding to the assumption that radiation energy consists of indivisible point-like localized quanta of energy hv [and of momentum hv/c, (c = velocity of light)], which are reflected undivided. This way of looking at the problem showed in a drastic and direct way that a type of immediate reality has to be ascribed to Planck's quanta, that radiation must, therefore, possess a kind of molecular structure as far as its energy is concerned, which of course contradicts Maxwell's theory. Considerations about radiation based directly on Boltzmann's entropy probability relation (probability taken to equal statistical temporal frequency) also lead to the same result. This dual nature of radiation (and of material corpuscles) is a major property of reality, which has been interpreted by quantum mechanics in an ingenious and amazingly successful fashion. This interpretation, which is looked upon as essentially definitive by almost all contemporary physicists, appears to me to be only a temporary expedient; a few remarks to this [point] will follow later.

Reflections of this type made it clear to me as long ago as shortly after 1900, i.e., shortly after Planck's trailblazing work, that neither mechanics nor electrodynamics could (except in limiting cases) claim exact validity. Gradually I despaired of the possibility of discovering the true laws by means of constructive efforts based on known facts. The longer and the more desperately I tried, the more I came to the conviction that only the discovery of a universal formal principle could lead us to assured results. The example I saw before me was thermodynamics. The general principle was there given in the theorem: The laws of nature are such that it is impossible to construct a *perpetuum mobile* (of the first and second kind). How, then, could such a universal principle be found? After ten years of reflection such a principle resulted from a paradox upon which I had already hit at the age of sixteen: If I pursue a beam of light with the velocity c (velocity of light in a vacuum), I should observe such a beam of light as an electromagnetic field at rest though spatially oscillating. There seems to be no such thing, however, neither on the basis of experience nor according to Maxwell's equations. From the very beginning it appeared to me intuitively clear that, judged from the standpoint of such an observer, everything would have to happen according to the same laws as for an observer who, relative to the earth, was at rest. For how should the first observer know, or be able to determine, that he is in a state of fast uniform motion?

One sees that in this paradox the germ of the special relativity theory is already contained. Today everyone knows, of course, that all attempts to clarify this paradox satisfactorily were condemned to failure as long as the axiom of the absolute character of time, or of simultaneity, was rooted unrecognized in the unconscious. To recognize clearly this axiom and its arbitrary character already implies the essentials of the solution of the problem. The type of critical reasoning required for the discovery of this central point was decisively furthered, in my case, especially by the reading of David Hume's and Ernst Mach's philosophical writings.

One had to understand clearly what the spatial coordinates and the time fixation of an event signified in physics. The physical interpretation of the spatial coordinates presupposed a rigid body of reference, which, moreover, had to be in a more or less definite state of motion (inertial system). In a given inertial system the coordinates denoted the results of certain measurements with rigid (stationary) rods. (One should always be aware that the presupposition of the existence in principle of rigid rods is a presupposition suggested by approximate experience but is, in principle, arbitrary.) With such an interpretation of the spatial coordinates the question of the validity of Euclidean geometry becomes a problem of physics.

If, then, one tries to interpret the time of an event analogously, one needs a means for the measurement of the difference in time (a periodic process, internally determined, and realized by a system of sufficiently small spatial extension). A clock at rest relative to the system of inertia defines a local time. The local times of all space points taken together are the "time," which belongs to the selected system of inertia, if a means is given to "set" these clocks relative to each other. One sees that *a priori* it is not at all necessary that the "times" thus defined in different inertial systems agree with one another. One would have noticed this long ago if, for the practical experience of everyday life, light did not present (because of the large value of *c*) the means for fixing an absolute simultaneity.

The presuppositions of the existence (in principle) of (ideal, or perfect) measuring rods and clocks are not independent of each other; a light signal that is reflected back and forth between the ends of a rigid rod constitutes an ideal clock, provided that the postulate of the constancy of the light velocity in vacuum does not lead to contradictions.

The above paradox may then be formulated as follows. According to the rules of connection, used in classical physics, between the spatial coordinates and the time of events in the transition from one inertial system to another, the two assumptions of

(1) the constancy of the light velocity
(2) the independence of the laws (thus especially also of the law of the constancy of the light velocity) from the choice of inertial system (principle of special relativity)

are mutually incompatible (despite the fact that both taken separately are based on experience).

The insight fundamental for the special theory of relativity is this: The assumptions (1) and (2) are compatible if relations of a new type ("Lorentz transformation") are postulated for the conversion of coordinates and times of events. With the given physical interpretation of coordinates and time, this is by no means merely a conventional step but implies certain hypotheses concerning the actual behavior of moving measuring rods and clocks, which can be experimentally confirmed or disproved.

The universal principle of the special theory of relativity is contained in the postulate: The laws of physics are invariant with respect to Lorentz transformations (for the transition from one inertial system to any other arbitrarily chosen inertial system). This is a restricting principle for natural laws, comparable to the restricting principle of the nonexistence of the *perpetuum mobile* that underlies thermodynamics.

First a remark concerning the relation of the theory to "four-dimensional space." It is a widespread error that the special theory of relativity is supposed to have, to a certain extent, first discovered or, at any rate, newly introduced, the four-dimensionality of the physical continuum. This, of course, is not the case. Classical mechanics, too, is based on the four-dimensional continuum of space and time. But in the four-dimensional continuum of classical physics the subspaces with constant time value have an absolute reality, independent of the choice of the frame of reference. Because of this, the four-dimensional continuum breaks down naturally into a three-dimensional and a one-dimensional (time), so that the four-dimensional point of view does not force itself upon one as *necessary*. The special theory of relativity, on the other hand, creates a formal dependence between the way in which the space coordinates on the one hand, and the time coordinates on the other, must enter into the natural laws.

Minkowski's important contribution to the theory lies in the following: Before Minkowski's investigation it was necessary to carry out a Lorentz transformation on a law in order to test its invariance under such transformations; but he succeeded in introducing a formalism so that the mathematical form of the law itself guarantees its invariance under Lorentz transformations. By creating a four-dimensional tensor calculus, he achieved the same thing for the four-dimensional space that the ordinary vector calculus achieves for the three spatial dimensions. He also showed that the Lorentz transformation (apart from a different algebraic sign due to the special character of time) is nothing but a rotation of the coordinate system in the four-dimensional space.

First, a critical remark concerning the theory as it is characterized above. It is striking that the theory (except for the four-dimensional space) introduces two kinds of physical things, i.e., (1) measuring rods and clocks, (2) all other things, e.g., the electromagnetic field, the material point, etc. This, in a certain sense, is inconsistent; strictly speaking, measuring rods and clocks should emerge as solutions of the basic equations (objects consisting of moving atomic configurations), not, as it were, as theoretically self-sufficient entities. The procedure justifies itself, however, because it was clear from the very beginning that the postulates of the theory are not strong enough to deduce from them equations for physical events sufficiently complete and sufficiently free from arbitrariness in order to base upon such a foundation a theory of measuring rods and clocks. If one did not wish to forego a physical interpretation of the coordinates in general (something that, in itself, would be possible), it was better to permit such inconsistency—with the obligation, however, of eliminating it at a later stage of the theory. But one must not legitimize the sin just described so as to imagine that distances are physical entities of a special type, intrinsically different from other physical variables ("reducing physics to geometry," etc.).

We now shall inquire into the insights of a definitive nature that physics owes to the special theory of relativity.

(1) There is no such thing as simultaneity of distant events; consequently, there is also no such thing as immediate action at a distance in the sense of Newtonian mechanics. Although the introduction of actions at a distance, which propagate at the speed of light, remains feasible according to this theory, it appears unnatural; for in such a theory there could be no reasonable expression for the principle of conservation of energy. It therefore appears unavoidable that physical reality must be described in terms of continuous functions in space. The material point, therefore, can hardly be retained as a basic concept of the theory.

(2) The principles of the conservation of linear momentum and of energy are fused into one single principle. The inert mass of an isolated system is identical with its energy, thus eliminating mass as an independent concept.

Remark. The speed of light c is one of the quantities that occurs in physical equations as a "universal constant." If, however, one introduces as the unit of time, instead of the second, the time in which light travels 1 cm, c no longer occurs in the equations. In this sense one could say that the constant c is only an *apparent* universal constant.

It is obvious and generally accepted that one could eliminate two more universal constants from physics by introducing, instead of the gram and the centimeter, properly chosen "natural" units (for example, mass and radius of the electron).

If one considers this done, then only "dimensionless" constants could occur in the basic equations of physics. Concerning such, I would like to state a proposition that at present cannot be based upon anything more than upon a faith in the simplicity, i.e., intelligibility, of nature: there are no *arbitrary* constants of this kind; that is to say, nature is so constituted that it is possible logically to lay down such strongly determined laws that within these laws only rationally completely determined constants occur (not constants, therefore, whose numerical value could be changed without destroying the theory).

The special theory of relativity owes its origin to Maxwell's equations of the electromagnetic field. Conversely, the latter can be grasped formally in satisfactory fashion only by way of the special theory of relativity. Maxwell's equations are the simplest Lorentz-invariant field equations that can be postulated for an antisymmetric tensor derived from a vector field. This in itself would be satisfactory, if we did not know from quantum phenomena that Maxwell's theory does not do justice to the energetic properties of radiation. But as to how Maxwell's theory would have to be modified in a natural fashion, for this even the special theory of relativity offers no adequate foothold. Also, to Mach's question: "how does it come about that inertial systems are physically distinguished above all other coordinate systems?" this theory offers no answer.

That the special theory of relativity is only the first step of a necessary development became completely clear to me only in my efforts to represent gravitation in the framework of this theory. In classical mechanics, interpreted in terms of the field, the potential of gravitation appears as a *scalar* field (the simplest theoretical possibility of a field with a single component). Such a scalar theory of the gravitational field can easily be made invariant under the group of Lorentz transformations. The following program appears natural, therefore: The total physical field consists of a scalar field (gravitation) and a vector field (electromagnetic field); later insights may eventually make necessary the introduction of still more complicated types of fields; but to begin with one did not need to bother about this.

The possibility of realization of this program was, however, in doubt from the very first, because the theory had to combine the following things:

(1) From the general considerations of special relativity theory it was clear that the inertial mass of a physical system increases with the total energy (therefore, e.g., with the kinetic energy).

(2) From very accurate experiments (especially from the torsion balance experiments of Eötvös) it was empirically known with very high accuracy that the gravitational mass of a body is exactly equal to its inertial mass.

It followed from (1) and (2) that the *weight* of a system depends in a precisely known manner on its total energy. If the theory did not accomplish this or could not do it naturally, it was to be rejected. The condition is most naturally expressed as follows: The acceleration of a system falling freely in a given gravitational field is independent of the nature of the falling system (especially therefore also of its energy content).

It turned out that, within the framework of the program sketched, this simple state of affairs could not at all, or at any rate not in any natural fashion, be represented in a satisfactory way. This convinced me that within the structure of the special theory of relativity there is no niche for a satisfactory theory of gravitation.

Now it came to me: the fact of the equality of inertial and gravitational mass, i.e., the fact of the independence of the gravitational acceleration from the nature of the falling substance, may be expressed as follows: In a gravitational field (of small spatial extension) things behave as they do in a space free of gravitation, if one introduces into it, in place of an "inertial system," a frame of reference accelerated relative to the former.

If then one interprets the behavior of a body with respect to the latter frame of reference as caused by a "real" (not merely apparent) gravitational field, it is possible to regard this frame as an "inertial system" with as much justification as the original reference system.

So, if one considers pervasive gravitational fields, not *a priori* restricted by spatial boundary conditions, physically possible, then the concept of "inertial system" becomes completely empty. The concept of "acceleration relative to space" then loses all meaning and with it the principle of inertia along with the paradox of Mach.

The fact of the equality of inertial and gravitational mass thus leads quite naturally to the recognition that the basic postulate of the special theory of relativity (invariance of the laws under Lorentz transformations) is too narrow, i.e., that an invariance of the laws must be postulated also relative to *nonlinear* transformations of the coordinates in the four-dimensional continuum.

This happened in 1908. Why were another seven years required for the construction of the general theory of relativity? The main reason lies in the fact that it is not so easy to free oneself from the idea that coordinates must have a direct metric significance. The transformation took place in approximately the following fashion.

We start with an empty, field-free space, as it occurs—related to an inertial system—within the meaning of the special theory of relativity, as the simplest of all imaginable physical situations. If we now think of a noninertial system introduced by assuming that the new system is uniformly accelerated against the inertial system (in a three-dimensional description) in one direction (conveniently defined), then there exists with reference to this system a static parallel gravitational field. The reference system may be chosen to be rigid, Euclidean in its three-dimensional metric properties. But the time in which the field appears as static is *not* measured by *equally constituted* stationary clocks. From this special example one can already recognize that the immediate metric significance of the coordinates is lost once one admits nonlinear transformations of the coordinates. To do the latter is, however, *obligatory* if one wants to do justice to the equality of gravitational and inertial mass through the foundations of the theory, and if one wants to overcome Mach's paradox regarding the inertial systems.

If, then, one must give up the notion of assigning to the coordinates an immediate metric meaning (differences of coordinates = measurable lengths, or times), one cannot but treat as equivalent all coordinate systems that can be created by the continuous transformations of the coordinates.

The general theory of relativity, accordingly, proceeds from the following principle: Natural laws are to be expressed by equations that are covariant under the group of continuous coordinate transformations. This group replaces the group of the Lorentz transformations of the special theory of relativity, which forms a subgroup of the former.

This postulate by itself is of course not sufficient to serve as point of departure for the derivation of the basic equations of physics. One might even deny, to begin with, that the postulate by itself involves a real restriction for the physical laws; for it will always be possible to reformulate a law, conjectured at first only for certain coordinate systems, so that the new formulation becomes formally generally covariant. Further, it is evident right away that an infinitely large number of field laws can be formulated that have this property of covariance. The eminent heuristic significance of the general principle of relativity is that it leads us to the search for those systems of equations that are *in their general covariant* formulation the *simplest ones possible*; among these we shall have to look for the field equations of physical space. Fields that can be transformed into each other by such transformations describe the same real situation.

The major question for anyone searching in this field is this: Of which mathematical type are the variables (functions of the coordinates) that permit the expression of the physical properties of space ("structure")? Only after that: Which equations are satisfied by those variables?

The answer to these questions is today by no means certain. The path chosen by the first formulation of the general theory of relativity can be characterized as follows. Even though we do not know by what kind of field variables (structure) physical space is to be characterized, we do know with certainty a special case: that of the "field-free" space in the special theory of relativity. Such a space is characterized by the fact that for a properly chosen coordinate system the expression

$$ds^2 = dx_1^2 + dx_2^2 + dx_3^2 - dx_4^2 \qquad (1)$$

belonging to two neighboring points, represents a measurable quantity (square of distance), and thus has a real physical meaning. Referred to an arbitrary system this quantity is expressed as follows:

$$ds^2 = g_{ik} dx_i dx_k \qquad (2)$$

whereby the indices run from 1 to 4. The g_{ik} form a (real) symmetrical tensor. If, after carrying out a transformation on field (1), the first derivatives of the g_{ik} with respect to the coordinates do not vanish, there exists a gravitational field with reference to this system of coordinates in the sense of the above consideration, but of a very special type. Thanks to Riemann's investigation of n-dimensional metric spaces, this special field can be characterized invariantly:

(1) Riemann's curvature-tensor R_{iklm}, formed from the coefficients of the metric (2), vanishes.
(2) The trajectory of a mass-point in reference to the inertial system (relative to which (1) is valid) is a straight line, hence an extremal (geodesic). This last statement, however, is already a characterization of the law of motion based on (2).

The *universal* law of physical space must be a generalization of the law just characterized. I now assumed that there are two steps of generalization:

(a) the pure gravitational field
(b) the general field (which is also to include quantities that somehow correspond to the electromagnetic field).

The case (a) was characterized by the fact that the field can still be represented by a Riemann metric (2), i.e., by a symmetric tensor, but without a representation of the form (1) (save on an infinitesimal scale). This means that in the case (a) the Riemann tensor does not vanish. It is clear, however, that in this case a field law must hold that is some generalization (loosening) of this law. If this generalized law also is to be of the second order of differentiation and linear in the second derivatives, then only the equation obtained by a single contraction

$$0 = R_{kl} = g^{im} R_{iklm}$$

was a prospective field law in the case (a). It appears natural, moreover, to assume that also in the case (a) the geodesic line is still to represent the law of motion of the material point.

It seemed hopeless to me at that time to venture the attempt of representing the total field (b) and to ascertain field laws for it. I preferred, therefore, to set up a preliminary formal frame for the representation of the entire physical reality; this was necessary in order to be able to investigate, at least preliminarily, the effectiveness of the basic idea of general relativity. This was done as follows.

In Newton's theory one can write the field law of gravitation thus:

$$\nabla^2 \phi = 0$$

(ϕ = gravitation potential), valid wherever the density of matter, ρ, vanishes. In general one has (Poisson's equation)

$$\nabla^2 \phi = 4\pi k \rho \ (\rho = \text{mass density}).$$

In the relativistic theory of the gravitational field, R_{ik} takes the place of $\nabla^2 \phi$. On the right-hand side we shall then have to replace ρ also by a tensor. Since we know from the special theory of relativity that the (inertial) mass equals the energy, we shall have to put on the right-hand side the tensor of energy density—more precisely, of the entire energy density that does not belong to the pure gravitational field. In this way one arrives at the field equation

$$R_{ik} - \tfrac{1}{2} g_{ik} R = - k T_{ik}.$$

The second member on the left-hand side is added because of formal considerations; for the left-hand side is written in such a way that its divergence, in the sense of the absolute differential calculus, vanishes identically. The right-hand side is a formal condensation of all things whose comprehension in the sense of a field theory is still problematic. Not for a moment, of course, did I doubt that this formulation was merely a makeshift in order to give the general principle of relativity a preliminary closed-form expression. For it was essentially *no more* than a theory of the gravitational field, which was isolated somewhat artificially from a total field of as yet unknown structure.

If anything in the theory as sketched—apart from the postulate of invariance of the equations under the group of continuous coordinate transformations—can possibly be claimed to be definitive, then it is the theory of the limiting case of a pure gravitational field and its relation to the metric structure of space. For this reason, in what immediately follows we shall speak only of the equations of the pure gravitational field.

The peculiarity of these equations lies, on the one hand, in their complicated structure, especially their nonlinear character with respect to the field variables and their derivatives, and, on the other hand, in the almost compelling necessity with which the transformation group determines this complicated field law. If one had stopped with the special theory of relativity, i.e., with the invariance under the Lorentz group, then the field law $R_{ik} = 0$ would remain invariant also within the frame of this narrower group. But, from the point of view of the narrower group, there would be no off-hand grounds for representing gravitation by a structure as involved as the symmetric tensor g_{ik}. If, nonetheless, one would find sufficient reasons for it, there would then arise an immense number of field laws out of quantities g_{ik}, all of which are covariant under Lorentz transformations (not, however, under the general group). Even if, however, of all the conceivable Lorentz-invariant laws, one had accidentally guessed precisely the law belonging to the wider group, one would still not have achieved the level of understanding corresponding to the general principle of relativity. For, from the standpoint of the Lorentz group, two solutions would incorrectly have to be viewed as physically different if they can be transformed into each other by a nonlinear transformation of coordinates, i.e., if from the point of view of the wider group they are merely different representations of the same field.

One more general remark concerning structure and group. It is clear that in general one will judge a theory to be the more nearly perfect the simpler a "structure" it postulates and the broader the group concerning which the field equations are invariant. One sees now that these two desiderata get in each other's way. For example: according to the special theory of relativity (Lorentz group) one can set up a covariant law for the simplest structure imaginable (a scalar field), whereas in the general theory of relativity (wider group of the continuous transformations of coordinates) there is an invariant field law only for the more complicated structure of the symmetric tensor. We have already given *physical* reasons for the fact that in physics invariance under the wider group has to be required:[2] from a purely mathematical standpoint I can see no necessity for sacrificing the simpler structure to the generality of the group.

The group of general relativity is the first one requiring that the simplest invariant law be no longer linear and homogeneous in the field variables and their derivatives. This is of fundamental importance for the following reason. If the field law is linear (and

2 To remain with the narrower group and at the same time to base the relativity theory of gravitation upon the more complicated [tensor-] structure implies a naive inconsequence. Sin remains sin, even if it is committed by otherwise ever so respectable men.

homogeneous), then the sum of two solutions is again a solution; so it is, for example, in Maxwell's field equations for the vacuum. In such a theory it is impossible to deduce from the field equations alone an interaction between structures that separately represent solutions of the system. That is why all theories up to now required, in addition to the field equations, special equations for the motion of material bodies under the influence of the fields. In the relativistic theory of gravitation, it is true, the law of motion (geodesic line) was originally postulated independently in addition to the field law. Subsequently, though, it turned out that the law of motion need not (and must not) be assumed independently, but that it is already implicitly contained within the law of the gravitational field.

The essence of this truly involved situation can be visualized as follows: A single material point at rest will be represented by a gravitational field that is everywhere finite and regular, except where the material point is located: there the field has a singularity. If, however, one computes the field belonging to two material points at rest by integrating the field equations, then this field has in addition to the singularities at the positions of the material points a curve of singular points connecting the two points. It is possible, however, to stipulate a motion of the material points so that the gravitational field determined by them does not become singular anywhere except at the material points. These are precisely those motions described in first approximation by Newton's laws. One may say, therefore: The masses move in such fashion that the solution of the field equations is nowhere singular except at the mass points. This property of the gravitational equations is intimately connected with their nonlinearity, and this, in turn, results from the wider group of transformations.

Now it would of course be possible to object: If singularities are permitted at the locations of the material points, what justification is there for forbidding the occurrence of singularities elsewhere? This objection would be justified if the equations of gravitation were to be considered as equations of the total field. [Since this is not the case], however, one will have to say that the field of a material particle will differ the more from a *pure gravitational field* the closer one comes to the location of the particle. If one had the field equations of the total field, one would be compelled to demand that the particles themselves could be represented as solutions of the complete field equations that are free of irregularities everywhere. Only then would the general theory of relativity be a *complete* theory.

Before I enter upon the question of the completion of the general theory of relativity, I must take a stand with reference to the most successful physical theory of our period, viz., the statistical quantum theory, which assumed a consistent logical form about twenty-five years ago, (Schrödinger, Heisenberg, Dirac, Born). At present this is the only theory that permits a unitary grasp of experiences concerning the quantum character of micro-mechanical events. This theory, on the one hand, and the theory of relativity on the other, are both considered correct in a certain sense, although all efforts to fuse them into a single whole so far have not met with success. This is probably why among contemporary theoretical physicists there exist entirely differing opinions as to what the theoretical foundation of the physics of the future will look like. Will it be a field theory? Will it be in essence a statistical theory? I shall briefly indicate my own thoughts on this point.

Physics is an attempt conceptually to grasp reality as something that is considered to be independent of its being observed. In this sense one speaks of "physical reality." In pre-quantum physics there was no doubt as to how this was to be understood. In Newton's theory reality was determined by a material point in space and time, in Maxwell's theory by the field in space and time. In quantum mechanics the situation is less transparent. If one asks: does a Ψ-function of the quantum theory represent a real fact in the same sense as a material system of points or an electromagnetic field? one hesitates to reply with a simple "yes" or "no." Why? What the Ψ function (at a definite time) states, is this: What is the probability for finding a definite physical quantity q (or p) in a definite given interval if I measure it at time t? The probability is here to be viewed as an empirically determinable, and therefore certainly a "real" quantity, which I may determine if I create the same Ψ-function very often and each time perform a q-measurement. But what about the single measured value of q? Did the respective individual system have this q-value even before the measurement? To this question there is no definite answer within the framework of the [existing] theory, since the measurement is a process that implies a finite disturbance of the system from the outside; it would therefore be conceivable that the system obtains a definite numerical value for q (or p), the measured numerical value, only through the measurement itself. For the further discussion I shall assume two physicists, A and B, who represent different conceptions concerning the real situation as described by the Ψ-function.

A. The individual system (before the measurement) has a definite value of q (or p) for all variables of the system, specifically *that* value which is determined by a measurement of this variable. Proceeding from this conception, he will state: The Ψ-function is not a complete description of the exact state of the system, but only an incomplete representation; it expresses only what we know about the system because of previous measurements.

B. The individual system (before the measurement) has no definite value of q (or p). The measured value is produced by the act of measurement itself consistent with the probability appropriate to the Ψ-function. Proceeding from this conception, he will (or, at least, he may) state: The Ψ-function is an exhaustive description of the real situation of the system.

Now we present to these two physicists the following case. There is to be a system that at the time t of our observation consists of two component systems S_1 and S_2, which at this time are spatially separated and (in the sense of the classical physics) interact with each other but slightly. The total system is to be described completely in terms of quantum mechanics by a known Ψ-function, say Ψ_{12}. All quantum theoreticians now agree upon the following. If I make a complete measurement of S_1, I obtain from the results of the measurement and from Ψ_{12} an entirely definite Ψ-function Ψ_2 of the system S_2. The character of Ψ_2 then depends upon *what kind* of measurement I perform on S_1.

Now it appears to me that one may speak of the real state of the partial system S_2. To begin with, before performing the measurement on S_1, we know even less of this real state than we know of a system described by the Ψ-function. But on one assumption we should, in my opinion, insist without qualification: the real state of the system S_2 is independent of any manipulation of the system S_1, which is spatially separated from the former. According to the type of measurement I perform on S_1, I get, however, a very different Ψ_2 for the second partial system $(\Psi_2, \Psi_2^1, \ldots)$. . Now, however, the real state of S_2 must be independent of what happens to S_1. For the same real state of S_2 it is possible therefore to find (depending on one's choice of the measurement performed on S_1) different types of Ψ-function. (One can escape from this conclusion only by either assuming that the measurement of S_1 (telepathically) changes the real state of S_2 or by denying altogether that spatially separated entities possess independent real states. Both alternatives appear to me entirely unacceptable.)

If now the physicists A and B accept this reasoning as valid, then B will have to give up his position that the Ψ-function constitutes a complete description of a real state. For in this case it would be impossible that two different types of Ψ-functions could be assigned to the identical state of S_2.

The statistical character of the present theory would then follow necessarily from the incompleteness of the description of the systems in quantum mechanics, and there would no longer exist any ground for the assumption that a future foundation of physics must be based upon statistics.

It is my opinion that the contemporary quantum theory represents an optimal formulation of the relationships, given certain fixed basic concepts, which by and large have been taken from classical mechanics. I believe, however, that this theory offers no useful point of departure for future development. This is the point at which my expectation deviates most widely from that of contemporary physicists. They are convinced that it is impossible to account for the essential aspects of quantum phenomena (apparently discontinuous and temporally not determined changes of the state of a system, simultaneously corpuscular and undulatory qualities of the elementary carriers of energy) by means of a theory that describes the real state of things [objects] by continuous functions of space for which differential equations are valid. They are also of the opinion that in this way one cannot understand the atomic structure of matter and of radiation. They rather expect that systems of differential equations, which might be considered for such a theory, in any case would have no solutions that would be regular (free from singularities) everywhere in four-dimensional space. Above everything else, however, they believe that the apparently discontinuous character of elementary processes can be described only by means of an essentially statistical theory, in which the discontinuous changes of the systems are accounted for by continuous changes of the probabilities of the possible states.

All of these remarks seem to me to be quite impressive. But the crux of the matter appears to me to be this question: What can be attempted with some hope of success in view of the present situation of physical theory? Here it is the experiences with the theory of gravitation that determine my expectations. In my opinion, these equations are more likely to tell us something *precise* than all other equations of physics. Take, for instance, Maxwell's equations of empty space by way of comparison. These are formulations corresponding to our experiences with infinitely weak electromagnetic fields. This empirical origin already determines their linear form; it has, however, already been emphasized above that the true laws cannot be linear. Such linear laws fulfill the superposition principle for their solutions; hence they contain no assertions concerning the interaction of elementary bodies. The true laws cannot be linear, nor can they be derived from such. I have learned something else from the theory of gravitation: no collection of empirical facts however comprehensive can ever lead to the setting up of such complicated equations. A theory can be tested by experience, but there is no way from experience to the construction of a theory. Equations of such complexity as are the equations of the gravitational field can be found only through the discovery of a logically simple mathematical condition that determines the equations completely or almost completely. Once one has obtained those sufficiently strong formal conditions, one requires only little knowledge of facts for the construction of the theory; in the case of the equations of gravitation it is the four-dimensionality and the symmetric tensor as expression for the structure of space that, together with the invariance with respect to the continuous transformation group, determine the equations all but completely.

Our task is that of finding the field equations for the total field. The desired structure must be a generalization of the symmetric tensor. The group must not be any narrower than that of the continuous transformations of coordinates. If one introduces a richer structure, then the group will no longer determine the equations as strongly as in the case of the symmetrical tensor as structure. Therefore it would be most beautiful if one were to succeed in expanding the group once more in analogy to the step that led from special relativity to general relativity. More specifically, I have attempted to draw upon the group of the complex transformations of the coordinates. All such endeavors were unsuccessful. I also gave up an open or concealed increase in the number of dimensions of space, an endeavor originally undertaken by Kaluza that, with its projective variant, even today has its adherents. We shall limit ourselves to the four-dimensional space and to the group of the continuous real transformations of coordinates. After many years of fruitless searching, I consider the solution sketched in what follows the one that is logically most satisfying.

In place of the symmetric g_{ik} ($g_{ik} = g_{ki}$), the nonsymmetric tensor g_{ik} is introduced. This quantity is composed of a symmetric part s_{ik} and of a real or purely imaginary antisymmetric a_{ik}, thus:

$$g_{ik} = s_{ik} + a_{ik}.$$

Viewed from the standpoint of the group, the combination of s and a is arbitrary, because the tensors s and a individually have tensor character. It turns out, however, that these g_{ik} (viewed as a whole) play a quite analogous role in the construction of the new theory to the symmetric g_{ik} in the theory of the pure gravitational field.

This generalization of the space structure seems natural also from the standpoint of our physical knowledge, because we know that the electromagnetic field involves an antisymmetric tensor.

For the theory of gravitation it is furthermore essential that from the symmetric g_{ik} it is possible to form the scalar density $\sqrt{|g_{ik}|}$ as well as the contravariant tensor g_{ik} according to the definition

$$g_{ik}g^{il} = \delta_k^l \quad (\delta_k^l = \text{Kronecker tensor}).$$

These structures can be defined in precise correspondence for the nonsymmetric g_{ik}, including tensor densities.

In the theory of gravitation it is further essential that, for a given symmetric g_{ik}-field, a field Γ_{ik}^l can be defined, which is symmetric in the subscripts and which, considered geometrically, governs the parallel displacement of a vector. Analogously for the nonsymmetric g_{ik} a nonsymmetric Γ_{ik}^l can be defined, according to the formula

$$g_{ik,l} - g_{sk}\Gamma_{il}^s - g_{is}\Gamma_{lk}^s = 0, \tag{A}$$

which accords with the corresponding relation of the symmetric g, only that, of course, one must pay attention here to the position of the lower indices in the g and Γ.

Just as in the theory with symmetric g_{ik}, it is possible to form a curvature R^i_{klm} out of the Γ, and from it a contracted curvature R_{kl}. Finally, by employing a variational principle together with (A), it is possible to find compatible field equations:

$$g^{is},s = 0 \quad (g^{ik} = \tfrac{1}{2}(g^{ik} - g^{ki})\sqrt{-|g_{ik}|}) \tag{B_1}$$

$$\Gamma^s_{is} = 0 \quad (\Gamma^s_{is} = 1/2\,(\Gamma^s_{is} - \Gamma^s_{si})) \tag{B_2}$$

$$R_{\underline{ik}} = 0 \tag{C_1}$$

$$R_{\underline{kl},m} + R_{\underline{lm},k} + R_{\underline{mk},l} = 0 \tag{C_2}$$

Each of the two equations (B_1), (B_2) is a consequence of the other if (A) is satisfied. $R_{\underline{kl}}$ denotes the symmetric, $R_{\underset{\smile}{kl}}$ the antisymmetric part of R_{kl}.

If the antisymmetric part of g_{ik} vanishes, these formulas reduce to (A) and (C_1)—the case of the pure gravitational field.

I believe that these equations constitute the most natural generalization of the equations of gravitation.[3] The proof of their physical usefulness is a tremendously difficult task, inasmuch as mere approximations will not suffice. The question is: What solutions do these equations have that are regular everywhere?

This exposition has fulfilled its purpose if it shows the reader how the efforts of a life hang together and why they have led to expectations of a certain kind.

A. EINSTEIN
Institute for Advanced Study
Princeton, New Jersey
[ca. 1946]

---

3 The theory here proposed, according to my view, has a fair probability of being found valid, if the way to an exhaustive description of physical reality on the basis of the continuum turns out to be at all feasible.

REFERENCES

AEA. Albert Einstein Archives Online. Hebrew University of Jerusalem. http://www.alberteinstein.info.

Bacciagaluppi, Guido and Antony Valentini. 2009. *Quantum Theory at the Crossroads: Reconsidering the 1927 Solvay Conference*. Cambridge: Cambridge University Press.

Bell, J. S. 1987. *Speakable and Unspeakable in Quantum Mechanics*. Cambridge: Cambridge University Press.

Beller, Mara. 1999. *Quantum Dialogue: The Making of a Revolution*. Chicago: University of Chicago Press.

Blum, Alexander, Roberto Lalli, and Jürgen Renn. 2015. "The Reinvention of General Relativity: A Historiographical Framework for Assessing One Hundred Years of Curved Space-time," *Isis* 106 (3): 598–620.

———. 2016 "The Renaissance of General Relativity: How and Why It Happened," *Ann. Phys.* 528, no. 5: 344–349.

———. (guest editors). 2017. "The Renaissance of Einstein's Theory of Gravitation (special issue of *EPJ H*)," *EPJ H* 42.

———. 2018. "Gravitational Waves and the Long Relativity Revolution," *Nature Astronomy* 2: 534–543.

Born, Max. 1949/1970. "Einstein's Statistical Theories." In *Albert Einstein: Philosopher-Scientist*, ed. Paul Arthur Schillp. The Library of Living Philosophers, vol. 7, pp. 161–177. Evanston, IL: Library of Living Philosophers.

———. 2005. *The Born-Einstein Letters: Friendship, Politics, and Physics in Uncertain Times: Correspondence between Albert Einstein and Max and Hedwig Born from 1916 to 1955 with commentaries by Max Born*. New York: Macmillan.

Calaprice, Alice, ed. 2005. *The New Quotable Einstein*. Rev. ed. Princeton, NJ: Princeton University Press.

Canales, Jimena. 2015. *The Physicist and the Philosopher: Einstein, Bergson, and the Debate That Changed Our Understanding of Time*. Princeton, NJ: Princeton University Press.

Carus, Paul, ed. 1902. *Kant's Prolegomena to Any Future Metaphysics* (1783). Chicago: Open Court Publishing.

CPAE. The Collected Papers of Albert Einstein Online. https://einsteinpapers.press.princeton.edu.

CPAE, vol. 1: Stachel, John, David C. Cassidy, and Robert Schulmann, eds. 1987. *The Collected Papers of Albert Einstein*. Vol. 1, *The Early Years, 1879–1902*. English translation. Princeton, NJ: Princeton University Press.

CPAE, vol. 2: Stachel, John, David C. Cassidy, Jürgen Renn, and Robert Schulmann, eds. 1990. *The Collected Papers of Albert Einstein*. Vol. 2, *The Swiss Years: Writings, 1900–1909*. English translation. Princeton, NJ: Princeton University Press.

CPAE, vol. 5: Klein, Martin J., A. J. Kox, and Robert Schulmann, eds. 1994. *The Collected Papers of Albert Einstein*. Vol. 5, *The Swiss Years: Correspondence, 1902–1914*. English translation. Princeton, NJ: Princeton University Press.

CPAE, vol. 6: Kox, A. J., Martin J. Klein, and Robert Schulmann, eds. 1997. *The Collected Papers of Albert Einstein. Vol. 6, The Berlin Years: Writings, 1914–1917*. English translation. Princeton, NJ: Princeton University Press.

CPAE, vol. 7: Janssen, Michael, Robert Schulman, József Illy, Christoph Lehner, and Diana Kormos Buchwald, eds. 2002. *The Collected Papers of Albert Einstein*. Vol. 7, *The Berlin Years: Writings, 1918–1921*. English translation. Princeton, NJ: Princeton University Press.

CPAE, vol. 8: Schulmann, Robert, A. J. Kox, Michel Janssen, Józef Illy, eds. 1998. *The Collected Papers of Albert Einstein. Vol. 8, The Berlin Years: Correspondence, 1914–1918*. English translation. Princeton, NJ: Princeton University Press.

CPAE, vol. 12: Kormos Buchwald, Diana, Ze'ev Rosenkranz, Tilman Sauer, Jözsef Illy, and Virginia Iris Holmes, eds. 2009. *The Collected Papers of Albert Einstein*. Vol. 12, *The Berlin Years: Correspondence January–December 1921*. English translation. Princeton, NJ: Princeton University Press.

CPAE, vol. 14: Kormos Buchwald, Diana, József Illy, Ze'ev Rosenkranz, Tilman Sauer and Osik Moses, eds. 2015. *The Collected Papers of Albert Einstein*. Vol. 14, *The Berlin Years: Writings & Correspondence, April 1923–May 1925*. English translation. Princeton, NJ: Princeton University Press.

Dongen, Jeroen van. 2010. *Einstein's Unification*. Cambridge: Cambridge University Press.

Einstein, Albert. 1901. "Conclusions Drawn from the Phenomena of Capillarity," *Annalen der Physik* 4: 513–523. Reprinted in CPAE vol. 2, Doc. 1.

———. 1916. "Ernst Mach," *Physikalische Zeitschrift*, 17: 101–104. In CPAE vol. 6, Doc. 29.

———. 1934. *Mein Weltbild*. Amsterdam: Querido.

———. 1946. "Remarks on Bertrand Russell's Theory of Knowledge." In *The Philosophy of Bertrand Russell*, ed. Paul Arthur Schilpp, The Library of Living Philosophers, vol. 5. Evanston, IL: Library of Living Philosophers. Originally published 1944.

———. 1949/1970. "Reply to Criticisms." In *Albert Einstein: Philosopher-Scientist*, ed. Paul Arthur Schillp. The Library of Living Philosophers, vol. 7, pp. 663–688. Evanston, IL: Library of Living Philosophers.

———. 1954. *Ideas and Opinions: Based on "Mein Weltbild,"* ed. Carl Seelig. New York: Bonanza Books.

———. 1955. *The Meaning of Relativity*, 5th ed. Princeton, NJ: Princeton University Press.

———. 1987. *Letters to Solovine*. New York: Philosophical Library.

———. 2000 [1934]. *The World as I See It*. Translated by Alan Harris. New York: Citadel Press.

———. 2015. *Relativity: The Special and the General Theory; 100th Anniversary Edition*, ed. Hanoch Gutfreund and Jürgen Renn. Princeton NJ: Princeton University Press.

Einstein, Albert, and Max Born. *Briefwechsel, 1916–1955*. Munich, Nymphenburger, 1969.

Einstein, Albert and Jakob Grommer. 1927. "Allgemeine Relativitätstheorie und Bewegungsgesetz," *Sitzungsber. phys-math. Kl.* 1: 235–245.

Einstein, Albert and Marcel Grossmann. 1913. "Outline of a Generalized Theory of Relativity and of a Theory of Gravitation." In *The Collected Papers of Albert Einstein*, vol. 4, Doc. 13, pp. 151–188.

Einstein, Albert, Leopold Infeld, and Banesh Hoffmann. 1938. "The Gravitational Equations and the Problem of Motion," *Annals of Mathematics* 39: 65–100.

Einstein, Albert, Boris Podolsky, and Nathan Rosen. 1935. "Can Quantum-Mechanical Description of Physical Reality Be Considered Complete?" *Physical Review* 47, no. 10: 777–780.

Elkana, Yehuda. 2008. "Einstein and God." In *Einstein for the 21st Century: His Legacy in Science, Art, and Modern Culture*, ed. Peter L. Galison, Gerald Holton, and Silvan S. Schweber, 35–47. Princeton, NJ: Princeton University Press.

Engler, Fynn Ole, and Jürgen Renn. 2018. *Gespaltene Vernunft: Vom Ende eines Dialogs zwischen Wissenschaft und Philosophie*. Berlin: Mattes & Seitz.

Fraenkel, Abraham H. 1954. "The Intuitionistic Revolution in Mathematics and Logic," *Bulletin of the Research Council of Israel* 3: 283–289.

Frank, Philipp. 1947. *Einstein: His Life and Times*. New York: Alfred Knopf.

French, A. P., ed. 1979. *Einstein: A Centenary Volume*. Cambridge, MA: Harvard University Press.

Graf-Grossmann, Claudia E. 2018. *Marcel Grossmann: For the Love of Mathematics*. Cham, Switzerland: Springer.

Greenberger, Daniel, ed. 2009. *Compendium of Quantum Physics: Concepts, Experiments, History and Philosophy*. Berlin: Springer.

Gutfreund, Hanoch. 2015. "Zwei der Glänzendsten Gestirne: Max Planck und Albert Einstein." In *Berlins wilde Energien: Porträts aus der Geschichte der Leibnizschen Wissenschaftsakademie*, ed. S. Leibfried, C. Markschies, E. Osterkamp, G. Stock, 310–343. Berlin: De Gruyter Akademie Forschung.

Gutfreund, Hanoch, and Jürgen Renn. 2015. *The Road to Relativity: The History and Meaning of Einstein's "The Foundation of General Relativity," Featuring the Original Manuscript of Einstein's Masterpiece*. Princeton, NJ: Princeton University Press.

———. 2017. *The Formative Years of Relativity: The History and Meaning of Einstein's Princeton Lectures*. Princeton NJ: Princeton University Press.

Hadamard, Jacques. 1945. *A Mathematician's Mind*. Princeton, NJ: Princeton University Press.

Hentschel Klaus. 1990. *Interpretationen und Fehlinterpretationen der speziellen und allgemeinen Relativitaetstheorie durch Zeitgenossen Albert Einstein*. Basel: Birkhauser.

Hoffmann, Banesh and Helen Dukas. 1972. *Albert Einstein: Creator and Rebel*. New York: Plume Books.

Howard, Don. 2014. "Einstein and the Development of Twentieth-Century Philosophy of Science." In *The Cambridge Companion to Einstein*, ed. Michel Janssen and Christoph Lehner, pp. 354–376. Cambridge: Cambridge University Press.

Holton, Gerald. 1992. "Ernst Mach and the Fortunes of Positivism in America," *Isis* 83, no. 1: 27–60.

———. 1995. *Einstein, History, and Other Passions*. Woodbury, NY: American Institute of Physics Press, Press.

———. 2008. "Who Was Einstein? Why Is He Still So Alive?" In *Einstein for the 21st Century: His Legacy in Science, Art, and Modern Culture*, ed. Peter L. Galison, Gerald Holton, and Silvan S. Schweber, 3–14. Princeton, NJ: Princeton University Press.

Holton, Gerald, and Yehuda Elkana, eds. 1982. *Albert Einstein: Historical and Cultural Perspectives: The Centennial Symposium in Jerusalem*. Princeton, NJ: Princeton University Press.

Janssen, Michel and Christoph Lehner, eds. 2014. *The Cambridge Companion to Einstein*. Cambridge: Cambridge University Press.

Janssen, Michel and Jürgen Renn. 2015. "Arch and Scaffold: How Einstein Found His Field Equations," *Physics Today* 68: 30–36.

Jerome, Fred, and Rodger Taylor. 2005. *Einstein on Race and Racism*. New Brunswick, NJ: Rutgers University Press.

Klein, Martin. 1982. "Fluctuations and Statistical Physics in Einstein's Early Work." In *Albert Einstein: Historical and Cultural Perspectives*, ed. Gerald Holton and Yehuda Elkana. Princeton, NJ: Princeton University Press.

Kollross, Louis. 1955. "Erinnerungen eines Kommilitonen." In *Helle Zeiten—Dunkle Zeiten: In Memoriam Albert Einstein*, ed. Carl Seelig, 17–31. Zurich: Europa Verlag.

Lehmkuhl, Dennis. 2019. "General Relativity as a Hybrid Theory: The Genesis of Einstein's Work on the Problem of Motion," *Studies in History and Philosophy of Science Part B: Studies in History and Philosophy of Modern Physics*, 67: 176–190.

Lehner, Christoph. 2014. "Realism and Einstein's Critique of Quantum Mechanics." In *The Cambridge Companion to Einstein*, ed. Michel Janssen and Christoph Lehner, pp. 306–353. Cambridge: Cambridge University Press.

Margenau, Henry. 1949/1970. "Einstein's Conception of Reality." In *Albert Einstein: Philosopher-Scientist*, ed. Paul Arthur Schillp. The Library of Living Philosophers, vol. 7, pp. 243–268. Evanston, IL: Library of Living Philosophers.

Menger, Karl. 1949/1970. "Theory of Relativity and Geometry." In *Albert Einstein: Philosopher-Scientist*, ed. Paul Arthur Schillp. The Library of Living Philosophers, vol. 7, pp. 457–474. Evanston, IL: Library of Living Philosophers.

Nathan, Otto, and Heinz Norden, eds. 1960. *Einstein on Peace*. New York: Avenel Books.

Nobel Foundation, ed. 1967. *Nobel Lectures: Physics, 1901–1921*. Amsterdam: Elsevier.

Norton, John. 2014. "Einstein's Special Theory of Relativity and the Problems of Electrodynamics That Led Him to It." In *The Cambridge Companion to Einstein*, ed. Michel Janssen and Christoph Lehner, pp. 72–102. Cambridge: Cambridge University Press.

Nye, Mary Jo. 1972. *Molecular Reality: A Perspective on the Scientific Work of Jean Perrin*. London: MacDonald.

Planck, Max. 1934. *Wege zur Physikalischen Erkenntnis: Reden und Vorträge*. Leipzig: Hirzel.

———. 1949. *Scientific Autobiography and Other Papers*. New York: Philosophical Library.

Reichenbach, Hans. 1949/1970. "The Philosophical Significance of the Theory of Relativity." In *Albert Einstein: Philosopher-Scientist*, ed. Paul Arthur Schillp. The Library of Living Philosophers, vol. 7, pp. 287–311. Evanston, IL: Library of Living Philosophers.

Reiser, Anton [Rudolf Kayser]. 1930. *Einstein: A Biographical Portrait*. New York: Boni.

Renn, Jürgen. 1997. "Einstein's Controversy with Drude and the Origin of Statistical Mechanics," *Archive for History of Exact Sciences* 51, no. 4: 315–354.

———. 2013. "Einstein as a Missionary of Science." *Science & Education* 22: 2569–2591.

Renn, Jürgen and Robert Rynasiewicz. 2014. "Einstein's Copernican Revolution." In *The Cambridge Companion to Einstein*, ed. Michel Janssen and Christoph Lehner, pp. 38–71. Cambridge: Cambridge University Press.

Renn, Jürgen, and Robert Schulmann, eds. 1992. *Albert Einstein—Mileva Marić: The Love Letters*. Princeton, NJ: Princeton University Press.

Reves, Emery. 1945. *The Anatomy of Peace*. New York: Harper.

Rowe, David E., and Robert Schulmann. 2007. *Einstein on Politics: His Private Thoughts and Public Stands on Nationalism, Zionism, War, Peace, and the Bomb*. Princeton, NJ: Princeton University Press.

Sauer, Tilman. 2014. "Einstein's Unified Field Theory Program." In *The Cambridge Companion to Einstein*, ed. Michel Janssen and Christoph Lehner, pp. 281–305. Cambridge: Cambridge University Press.

Ryckman, Thomas. 2014. " 'A Believing Rationalist': Einstein and 'the Truly Valuable' in Kant." In *The Cambridge Companion to Einstein*, ed. Michel Janssen and Christoph Lehner, pp. 377–395. Cambridge: Cambridge University Press.

Schiller, F.S.C. 1934. *Must Philosophers Disagree? and Other Essays in Popular Philosophy*. London: Macmillan.

Schilpp, Paul Arthur, ed. 1944/1971. *The Philosophy of Bertrand Russell*. The Library of Living Philosophers, vol. 5. La Salle, IL: Open Court.

———, ed. 1949/1970. "Albert Einstein: Autobiographical Notes (in German, and in English Translation)." In *Albert Einstein: Philosopher-Scientist*. The Library of Living Philosophers, vol. 7, pp. 1–96. Evanston, IL: Library of Living Philosophers.

———, ed. 1979. *Albert Einstein: Autobiographical Notes*. La Salle, IL: Open Court.

Seelig, Carl. 1956. Helle Zeit—Dunkle Zeit: In Memoriam Albert Einstein. Zurich: Europa.

Sommerfeld, Arnold. 1949/1970. "To Albert Einstein's Seventieth Birthday." In *Albert Einstein: Philosopher-Scientist*, ed. Paul Arthur Schillp. The Library of Living Philosophers, vol. 7, pp. 97–105. Evanston, IL: Library of Living Philosophers.

Stachel, John. 1993. "The Other Einstein: Einstein Contra Field Theory," *Science in Context* 6, no. 1: 275–290.

———, ed. 2005. *Einstein's Miraculous Year: Five Papers That Changed the Face of Physics*. With a new introduction by John Stachel and a foreword by Roger Penrose. Princeton, NJ: Princeton University Press. Originally published 1998.

Straus, Ernst G. 1982. "Reminiscences." In *Albert Einstein: Historical and Cultural Perspectives*, ed. Gerald Holton and Yehuda Elkana. Princeton, NJ: Princeton University Press.

Vandenabeele, Bart. 2012. *A Companion to Schopenhauer*. Chichester: Blackwell Publishing.

INDEX

Page numbers in italics refer to boxes or figures.